"This truly inspiring and fascinating book leaves you never wanting to waste a single second ever again. Everything you need to know about how your brain works and how to maximize it is contained in an easy-to-read way. The book proves you really can do anything and there are lots of simple ways to help ensure you too can make the most of your biggest asset – your brain! Without doubt, a book you cannot be without!"
**Dame Sarah Storey, DBE**

"For all the debate about governments nudging people to make better decisions or to adopt better behaviours, it is easy to overlook the fact that we can actually nudge ourselves. This book is a wonderful guide to how to do just that."
**Rory Sutherland, Executive Creative Director and Vice Chairman, OgilvyOne London and Vice Chairman, Ogilvy Group UK**

"I thought it was accessible, thought-provoking and full of useful, easy-to-follow tips about improving your everyday life through a better understanding of the brain."
**Killian Fox, writer for *The Observer* and other publications**

"A really great book that explains in layman's terms how the brain works and how you can then translate that knowledge to enhance your own performance. Thought-provoking and insightful, it will add considerable value to anyone still willing to learn, irrespective of which rung of the success ladder they are on. It's an enjoyable and extremely useful read."
**Mark Hussein, Global Head of HSBC Commerical Insurance and Investments**

"*Sort Your Brain Out* is a must-read for everyone. It is a clever and thoughtful book designed to help the lay reader understand more about the brain's most intimate workings but most importantly it provides erudite yet easily consumed bite-sized gobbets of information on how to improve one's lobar lot. The fascinating examples are eminently readable and marvellously memorable; the reading of this

book will stretch the brain in exactly the way the authors have devised. This is mental stimulation at its best."
**Chantal Rickards, Head of Programming and Branded Content, MEC**

"As someone who has spent their life reviewing neuroscience material, I was struck by how the overview on offer contextualises some aspects of brain function in a novel and refreshing way.

In short, this is a delightful and illuminating read – it is the book that I would (will) give my family, when they ask searching questions about neuroscience – and what it means for them."
**Professor Karl Friston FRS, Scientific Director, Wellcome Trust Centre for Neuroimaging, University College London**

"*Sort Your Brain Out* is has clarity of purpose and many features that puts it ahead of its competitors in an expanding area of interest. Making the best use of the amazing brains we all inherit, even though they are destined to operate in a world far removed from the environment that shaped their evolution, is crucial. There probably is no more important a task for us as individuals or for the groups we live and work in than this. Help and the chance to expand our insight is at hand."
**Ian Edwards, Head of Strategy, Advertising Planning firm Vizeum**

"Engaging, accessible, demystifying."
**Dr Daniel Glaser, Director, Science Gallery London**

"This book explores the kind of topics we all think and talk about: Is the internet making us stupid? What do alcohol and caffeine really do to our brains? It provides you with exactly the kind of fascinating nuggets of information you end up reading out to whoever you happen to be with, as well as practical tips on how to maximise what we all have between our ears. Forget brainstorming, it's all about brainshaking and dunking now. Neuroscience demystified and simplified without being patronising; a must-read."
**Olivia Walmsley, *Mail Online***

# SORT YOUR BRAIN OUT

## Boost your performance, manage stress and achieve more

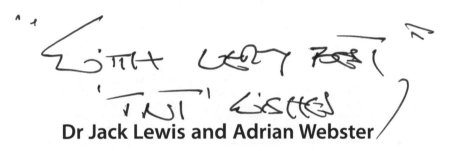

**Dr Jack Lewis and Adrian Webster**

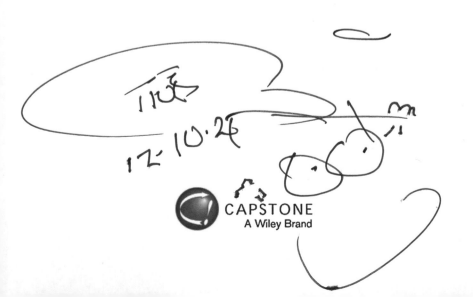

CAPSTONE
A Wiley Brand

This edition first published 2014
© 2014 Jack Lewis and Adrian Webster

*Registered office*
John Wiley and Sons Ltd, The Atrium, Southern Gate, Chichester, West Sussex, PO19 8SQ, United Kingdom

For details of our global editorial offices, for customer services and for information about how to apply for permission to reuse the copyright material in this book please see our website at www.wiley.com.

The right of the authors to be identified as the authors of this work has been asserted in accordance with the Copyright, Designs and Patents Act 1988.

Reprinted July 2014, August 2014, September 2014, October 2014, June 2015

Wiley publishes in a variety of print and electronic formats and by print-on-demand. Some material included with standard print versions of this book may not be included in e-books or in print-on-demand. If this book refers to media such as a CD or DVD that is not included in the version you purchased, you may download this material from http://booksupport.wiley.com. For more information about Wiley products, visit www.wiley.com.

Designations used by companies to distinguish their products are often claimed as trademarks. All brand names and product names used in this book and on its cover are trade names, service marks, trademark or registered trademarks of their respective owners. The publisher and the book are not associated with any product or vendor mentioned in this book. None of the companies referenced within the book have endorsed the book.

Limit of Liability/Disclaimer of Warranty: While the publisher and author have used their best efforts in preparing this book, they make no representations or warranties with the respect to the accuracy or completeness of the contents of this book and specifically disclaim any implied warranties of merchantability or fitness for a particular purpose. It is sold on the understanding that the publisher is not engaged in rendering professional services and neither the publisher nor the author shall be liable for damages arising herefrom. If professional advice or other expert assistance is required, the services of a competent professional should be sought.

*Library of congress cataloging-in-publication data*

Lewis, Jack,
    Sort your brain out / Dr Jack Lewis and Adrian Webster.
        pages cm
    Includes index.
    ISBN 978-0-857-08537-5 (pbk.)—ISBN 978-0-857-08536-8 (ebk)—ISBN 978-0-857-08535-1 (ebk)  1. Brain—Popular works.  2. Neurosciences.  I. Webster, Adrian,  II. Title.
    QP376.L47 2014
    612.8'2—dc23                                    2013048672

A catalogue record for this book is available from the British Library.

Illustrations by Charlie Johnson
Cover design by Mackerel

ISBN 978-0-857-08537-5 (paperback)    ISBN 978-0-857-08535-1 (ebk)
ISBN 978-0-857-08536-8 (ebk)

Set in 11/14 pt MyriadPro-Regular by Toppan Best-set Premedia Limited, Hong Kong
Printed in Great Britain by TJ International Ltd, Padstow, Cornwall, UK

In memory of Susan Rose McColl and
all those who were told they can't – but have.

# Contents

# What This Book Is All About

Every human being on this planet has the most incredible device in the known universe residing within their skull. Yet despite all of us having more or less the same make and model, the vast majority of users are completely unaware of its stunning capabilities; let alone its phenomenal ability to adapt.

The human brain adapts to the demands of pretty much any environment. It physically changes its own circuitry to improve performance in any behaviour that is regularly required of it. It will change in a manner that enables skills and abilities to become faster, more accurate and more efficient the next time you come to do them.

This process of rewiring for self-enhancement is so gradual that the day-to-day improvements are usually imperceptible. Only if you continue to perform that behaviour intensively, regularly and consistently – over an extended period of weeks and months – will your brain change sufficiently for the improvement in ability to become noticeable.

But brains don't only adapt to accommodate good behaviours, they adapt to perform *any* regularly repeated behaviour; without even thinking about it. Whether it is something useful like safely steering a car whilst your attention is consumed by an absorbing radio show, or not-so-helpful like helping yourself to that second slice of cake, your behaviour is largely controlled by a brain operating on autopilot, for better or for worse.

The aim of this book is to inspire you to consider the tremendous impact that neuroplasticity – your brain's ability to physically change to deal with any circumstances – can have on your behaviour. Current beliefs and behaviours were formed according to past events and notions regularly encountered in your daily life prior to now. Future beliefs and behaviours will be formed according to whatever thoughts, people and places you *choose* to regularly and intensely engage with from here on. Your brain is constantly upgrading and downgrading its circuits inside your head, every single day – for better or for worse.

By giving people a clearer understanding of how their own brains work and by changing the common but false perception that you "can't teach an old dog new tricks," the main objective of this book is to harness the revelation that, throughout adulthood, we can fundamentally change the very fabric of our brains. As a result, we can subtly alter our habitual behaviours, beliefs, motivations and, eventually, bring about profound positive change. This book will provide you with a wide range of simple brain-enhancing tools and practical tips to do just that.

# Introduction

**M**any people have a burning desire to be successful, some even have the know-how; and for those who don't there are literally thousands of self-help books out there, telling them how.

Yet, despite tsunamis of ambition and an abundant supply of well-meant advice, only a few people achieve real success and even fewer manage to maintain it.

The main reason is that despite being the most sophisticated piece of bio-wetware in the known universe, capable of running the most phenomenally complex software, your brain doesn't come with a user guide.

Incredibly, billions of people spend their lives scurrying around, all revved up, trying to get somewhere and devouring volumes of information on self-improvement. But they don't have a clue about the engine under their bonnet, its profound capabilities, or how best to use it.

In other words, they – as captain of the ship – may have all the drive, passion and heart's desire in the world to set and maintain a particular course, but if the engine room can't deliver, they'll be left wanting, drifting in the doldrums of success.

Most take their brain for granted, some even forget it exists, others spend hours at the gym working away on the bodywork, but only a few realize – that with a basic understanding and a small amount of care – just how much more they can get out of themselves.

In this book, we will help you to get a better understanding of how our brains work and explore ways of consistently getting more out of them. The eye-opening findings will explain the basic needs of our own – often idling but potentially brilliant – high-performance engines and, hopefully, help you achieve more from yours.

## A bit about us

We first met in January 2011 when we were both speaking at a conference in Tenerife. The theme of the event was "Are You Ready?" Our task was not only to inspire those attending but also to offer useful guidance to help them as a team be prepared for the tough challenges that lay ahead and enable them to capitalize on any opportunities heading their way.

As two very different people we found ourselves working together to deliver the same messages but from completely different angles. It was then that we realized just how impactful our combined knowledge could be, and what a difference it could make.

### Where Adrian's coming from

As a motivational business speaker I'd like to think that I am a highly motivated person, I'd be in the wrong job if I wasn't! I'd also like to think, having written self-help books, that I have a fairly good idea of what it takes to be successful.

However, despite being a reasonably fit, fairly intelligent and relatively successful person with bags of self drive and years of practical experience, especially when it comes to developing winning attitudes, I was keen to find out more about the hardware that supports the software – the engine that runs my mind.

I wanted to know more about my own brain, learn how to help it be even more productive and hopefully keep it in full working order for many more years to come.

Like you, I live in the real world. I run around at what often feels like a thousand miles an hour, juggling family, work and social commitments. There are times when, even as a motivational speaker, I feel a bit run down – especially when doing a lot of travelling. The gym can at times be very unappealing; and with my batteries running low I don't always feel as mentally sharp as I'd like to be.

As a writer I sometimes find it hard to be as creative as I know I can be, and despite having clear goals it's difficult to consistently stay focused. On top of all this, when I do get to spend time with the most important people in my life, my family, it can be a struggle some days to unwind; my overworked brain just doesn't seem to want to stop revving!

As an everyday person I wasn't under any illusion that overnight I'd suddenly gain the combined planet-sized intellectual skills of a mathematical genius, the creativity of a Renaissance master and the single-mindedness of an Olympic athlete. I just wanted to sharpen up a little, consistently have more energy, hopefully stay focused for longer, be a touch more creative and enjoy quality time with my family. At the end of the day, I just wanted to make the most of the one I've got.

As a lifelong learner I'm not ashamed to accept all the help I can get, so I decided to team up with TV's favourite neuroscientist, Dr Jack Lewis, to see just how much of an improvement I could make to my own brain. I'm pleased to report that his practical advice has had an extremely positive effect, and I have already noticed a tangible difference in my brain's performance.

As we progress through this book together and look at ways of optimizing the capabilities of brains, Jack and I are going to share with you all the practical advice that he had to offer me and, at the same time, draw on our diverse experiences to give you some helpful suggestions about how you too could improve the performance of your brain. Hopefully you'll take them on board, start using them and see what a difference they make to you.

*For more information about Adrian – please visit: www.adrianwebster.com or tweet @polarbearpirate*

## Where Jack's coming from

At school I found biology absolutely fascinating and pursued this passion into the realms of neurobiology. I eventually ended up in Germany where I carried out post-doctoral research on the edge of the Black Forest, using cutting-edge fMRI (functional Magnetic Resonance Imaging) scanning technology to plumb the depths of the human brain. Yet, since starting my PhD at University College London five years earlier, I had begun to feel increasingly frustrated. The neuroscience literature was brimming with fascinating revelations about the mysterious organ between our ears, and it seemed clear to me that these could genuinely help everyday people to better understand their own behaviour and that of those around them. But the only people that appeared to be reading it were geeks like me, and I felt that the rest of the world was missing out on something really valuable.

Thousands of experiments were accumulating every year, providing snapshots of how our brains sense the world around us, think, feel and make decisions – in a manner that, throughout human history, was previously unknown or completely misunderstood. All of these invaluable insights into our humanity were locked up inside pay-per-view academic journals from which only rigorously trained researchers could extract any meaning from the dense scientific jargon.

I have presented brain-related insights to millions of viewers across the world via the BBC, ITV, Channel 4, Sky, National Geographic, Discovery, TLC, Teacher's TV and even MTV! I have always had a burning desire to write a book that explains to "everyday" people (that's what I call non-scientists) how their brain works but I'd never found the right time to do this. That is until I spoke at the conference in Tenerife and the world of neuroscience merged with the world of motivation to cut through complex theory, and pass on compelling, much needed information to *everyday* people.

*For more information about Jack – please visit:*
*www.drjack.co.uk or tweet @drjacklewis*

# Your Amazing Brain

The word "amazing" seems to be used pretty loosely these days to describe a lot of things, many of which often turn out to be disappointingly mediocre; but in the case of your brain there is no other word that comes close to describing it.

This pink coloured, wrinkled lump of pulsating wetware, with a texture not dissimilar to blancmange, is composed of around 85% fat and weighs in at around 1.5 kg. It contains a densely woven meshwork of 86 billion brain wires along with a further 86 billion support cells, all neatly packed away between your ears. It is *truly* amazing.

As the ultimate supercomputer, your brain is currently light years ahead of anything that man has so far managed to create. It works relentlessly, non-stop around the clock, continuously reshaping to adapt our skills and behaviours to suit an almost infinite variety of different real and potential future circumstances. Receiving and delivering data, analyzing information, performing complex, multifunctional tasks in parallel and monitoring billions of functions; all at breathtaking speed. Its capabilities are quite staggering.

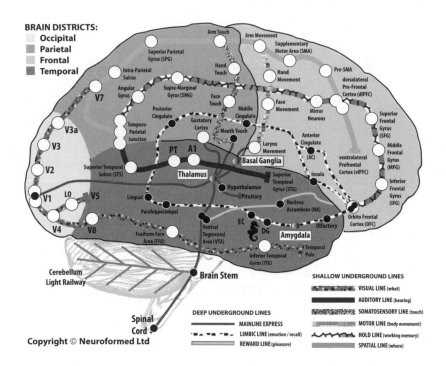

## But when it comes to performance, what does your brain look like?

The map in the illustration above shows some of the main stops on the underground system that is your very own brain. There would be no benefit in overloading you with unnecessary information by talking about every area of your brain, but it would be useful to start by pointing out three key areas that are most relevant to what we'll be discussing in this book.

You may be wondering why there is a seahorse in the illustration. The Hippocampus includes the DG (Dentate Gyrus) and EC (Entorhinal Cortex) tube stops on the lower part of the Limbic Line, a particularly dense area of networked brain wires that is interconnected with virtually every other part of your brain.

This part of your brain plays three key roles:

1. It helps you keep track of where you are in space – a basic GPS system that gives you a sense of where you are and how to get where you're going.

2. It enables you to create and recall memories of events and pieces of information.

3. It's even vital for the ability to imagine the future!

These two functions are closely related, as many of our memories of life events are closely intertwined with the places in which they were experienced. That way, when you return to that place, the relevant memories are triggered. Hence a visit to your old primary school can trigger a surge of long forgotten memories. In reality the Hippocampus cluster of tube stops in your brain is deep under the surface, at the core of your Temporal Lobe, which runs from behind your ear to behind your temples (hence the name).

## Why a seahorse?

If your brain's Hippocampus was surgically removed from the core of your Temporal Lobe, it would actually look very much like a seahorse; indeed Hippocampus actually means "horse" (hippo) – "sea monster" (campus) in ancient Greek.

Just to the right of the DG you'll find the Amygdala tube stop. This ever-alert brain area is responsible for, amongst other roles, generating various emotions, and constantly monitoring the sensory information spilling in from your surroundings for signs of potential danger. Like a military listening post for your brain, it is forever looking out for possible serious threats, always primed and ready to push the "fear response button" the split second one is detected. This is the part of your brain that, within less than a crotchet of time, having heard a loud bang or having spotted a rapidly approaching, incoming object causes you to freeze in your tracks or duck out of the way – before you are even fully aware of it. With your heart now pounding and your muscles flooded with blood, you're all set and ready for a confrontation or a hasty exit.

Just above the Amygdala tube stop is the Reward Line that passes deep through the very centre of your brain. It evolved to produce pleasure whenever you engage in behaviours that promote the survival of our species i.e. eating, drinking and having sex. Known collectively as the reward pathways – the VTA (or Ventral Tegmental Area), NA (Nucleus Accumbens) and OFC (Orbitofrontal Cortex) stops are also critical to decision making. As well as enabling us to feel pleasure in any given moment, the NA stop provides a reward-based prediction built around how much pleasure or benefit is likely to be derived from choosing one particular option over another. This means that not only are they instrumental in directing every single decision we ever make, they are also fundamental to the learning process. Without the reward system we would never learn from any of our mistakes.

To help give you a clearer perspective of what we're looking at here, the London Underground can, at this moment in time, proudly boast a combined track length of 250 miles, with tube trains travelling around between 270 stations at a top speed of up to 70 mph.

There are more connections between brain wires in your head than there are stars in our Galaxy – 0.15 quadrillion.

If all your brain wires were laid out end to end, they would be approximately 100,000 miles in length with hundreds of thousands of trillions of trains all travelling up and down, bang on time, at up to 250 miles per hour between 1,000 trillion connections; or synapses as they are known to science. And, if all these wires – the white matter of your brain – were laid out as an underground train network it would cover an area of 561,476 square miles, a surface area greater than that of South Africa, all tucked into a space smaller than your average pumpkin.

*But what really makes the human brain so very special is NEUROPLASTICITY. Its ability to continuously change, learn and, perhaps even more importantly, its ability to adapt to unexpected and widely varying circumstances in new and creative ways.*

Your brain can send up to one hundred, thousand, trillion messages per second using only the same amount of power as your average fridge light bulb. For a man-made supercomputer to send and receive that many messages per second it would require its own small hydroelectric power plant to provide the 10,000,000 watts needed to power it. Less than a litre of blood passing every minute through the brain of chess grandmaster Garry Kasparov was sufficient to keep his forehead cool to the touch, whilst his opponent – the IBM super computer Deep Blue – needed a fan-driven cooling system to prevent it from blowing up.

## Taxi!

Fresh challenges bring about physical changes to your brain. The drivers of London's famous black cabs spend years learning "The Knowledge," a seemingly indigestible mountain of information to commit to memory by anyone's standards. Included in it are the whereabouts of 25,000 streets along with 20,000 places of interest that at any time a fare-paying passenger, having hopped into the back of their cab, may ask to go to. During this period of exhaustive information ingestion, the Hippocampi of these determined wannabe cabbies physically grow larger due to all of the extra connections required to retain all that information – only to return to their normal size after retirement. It really is a case of use it or lose it!

What this shows is that your brain not only adapts to take on new challenges, it physically restructures itself to meet them. As yet, there is no computer capable of reconfiguring in this way to cope with new demands asked of it. Not bad for a design that first appeared on the scene back in the Stone Age, and which still out-competes the most complex systems of the modern age – well, for the moment anyway!

When we sleep our brain cells shrink to allow cerebrospinal fluid to seep into the gaps to wash away the metabolic waste that accumulates each and every day.

It doesn't come with a guarantee or any warranties, but if you look after your brain, it should remain fully functional and in good working order throughout your entire lifetime. And, if you're ever worried about running out of memory space, please don't! You'll be relieved to hear that it comes with the equivalent memory space of a one million gigabyte chip. That's enough memory to record over three million hours' worth of your favourite TV programmes.

Your brain is a phenomenal, unimaginably brilliant piece of kit and, please note, the emphasis on *your* brain – we all have the same make and model. Provided you are of this planet, that you are a human being, and your name isn't Albert Einstein, there would have been very little difference between the brain that you had and the brain the person sat next to you had when you both started out in life.

During early pregnancy 250,000 new neurons are born in the fetal brain every minute.

Yours may be the same make and model but when it comes to shaping your brain and differentiating it's individual performance from that of others, there are three very big influencing factors:

1. The environments that you spend most of your time interacting with

2. What you are exposed to in those environments

3. What your time in those environments is actually spent doing

Yes, our brains are all amazing but it is how we have made use of our brains over our lifetime so far that makes them so very different. More importantly, when it comes down to performance, it's what we now choose to do with them from here on that will determine just how well they continue to serve us in dealing with the daily demands of our own lifestyle.

# Flying Start

There are a few things that you could start doing right now to instantly improve your brain's performance and get you off to a flying start. Let's call them Brain Optimization Tips, or BOPs for short. You can follow these simple suggestions to get your brain firing on all cylinders each and every day. Here are five for starters:

## BOP1: Water – Start every single day by rehydrating your brain

Believe it or not your brain is 73% water. The efficiency with which it can send electrical messages around the 100,000 miles of brain wires is compromised when you are dehydrated. The trouble is you wake up slightly dehydrated every single day. How do we know this? Well, if you managed to survive the night then you can't have stopped breathing. The reason that breathing is absolutely vital when it comes to staying alive is because this is the only way you can get oxygen into your body and carbon dioxide out. In order for your lungs to get these gases moving in and out of your bloodstream successfully they must be moist. But that means that upon every exhalation, we release a little bit of water vapour. This of course happens all day and night, but during the daytime we replace the lost water whenever hunger or thirst motivates us to eat or drink. During the night there are fewer, if any, opportunities to replace this lost water; and so by morning time there is inevitably an imbalance to correct.

*BOP1: Drink a small glass of water the moment you wake up in the morning and make sure you stay well hydrated throughout the day – for your brain's sake.*

## BOP2: Exercise – Vital for brain health (and holding onto your marbles)

Exercise is well known to be good for the body but everyone overlooks its immense impact on the health of your brain. In the short term, the moment you start to do any form of even moderately demanding exercise, your body automatically responds by releasing a torrent of hormones and brain chemicals that make you feel good. Even more importantly, an ever-increasing body of evidence suggests that people who take regular exercise enjoy better brain function for longer. It actually increases the rate at which new brain cells are created in the Hippocampus. It's more important than any other factor in helping people to hold onto their marbles well into old age.

*BOP2: Do a minimum of 20 mins moderate to intense exercise, every other day.*

## BOP3: Stress – Control your cortisol to manage your stress

Cortisol is an extremely important hormone that is released in your body in response to the problems of everyday life. It is responsible for making you feel "stressed out," yet it is most definitely a friend and not a foe. We can all be agreed on the fact that feeling stressed is certainly not one of life's more pleasant experiences. But nonetheless, cortisol is a vital part of any happy, successful life. It helps us to get things done.

If cortisol did not make us feel uncomfortable we would lack the impetus to act. Not only that, but cortisol actually mobilizes body and brain to deal with life's daily stresses. It increases metabolism so that more physical and mental energy is available to deal with the problem. It puts the brain in an uncomfortable emotional state because we humans are fundamentally lazy – working hard only to remove discomfort and/or to gain greater comfort.

There is a natural daily rhythm to the release of cortisol ensuring that we are keyed up during the daytime and winding down towards bedtime. Bad news triggers a boost of cortisol to mobilize the extra resources required to help eliminate the problem.

So, stress is good but *chronic* stress is most definitely bad. This is because in order to help you deal with life's major challenges cortisol suppresses the immune system. This enables us to postpone feeling ill – which forces us to rest, directing energy and resources into fighting off the bugs – until a stressful situation has passed by or been resolved. Chronic stress describes a situation where cortisol levels remain high for many weeks and months, meaning that body and brain never get the chance to repair and fight disease. Whilst it is virtually impossible to remove the sources of stress in life there are several things you can do to reduce cortisol levels using the power of thought alone.

*BOP3: Manage stress by proactively setting aside GOM Time (see All Aboard the Stress Express! chapter), clinically proven to reduce cortisol levels.*

## BOP4: Sunshine – Soak up the sun's rays to stabilize your mood

When ultraviolet (UV) light strikes the skin the human body can make vitamin D. Vitamin D is used in the brain to make a very important brain chemical called serotonin. Serotonin is crucial because it is involved in a wide variety of brain pathways including those involved in mood regulation. The most widely prescribed class of medications for depression are those that increase levels of serotonin. For instance, SSRIs (selective serotonin reuptake inhibitors) like Prozac work by slowing down the rate at which any serotonin that has been released into your brain's synapses gets removed. By keeping serotonin levels up, mood improvement is also kept on the up. Ecstasy, aka MDMA (3,4-methylenedioxy-N-methamphetamine), is very popular amongst recreational drug users, and also increases levels of serotonin in the brain. However, rather than just blocking the reuptake of serotonin, like SSRIs do, it also triggers a massive increase in the release of serotonin into the synapse (a connection between one brain wire and the next) – leading to feelings of intense pleasure and empathy with others.

 Your skin weighs three times as much as your brain.

Rather than meddling with the release and/or reuptake of serotonin, getting your skin in direct sunlight every day will ensure there's a plentiful supply of serotonin readily available for your brain to regulate your mood more effectively. Bear in mind that your motivation to go out of your way to do this should fluctuate with the seasons. As the days get shorter the likelihood that you will not be getting enough UV to keep your serotonin levels topped up increases and so you should be more vigilant in getting into the daylight. Don't be put off by cloudy days, although cloud cover cuts out a large proportion of the visible light, making it seem a bit gloomy compared to when the sun is out, the strength of UV light is not diminished to the same degree. This is why you can still get a nasty sunburn on the beach even when it is overcast.

*BOP4: Get your skin in daylight for at least 5–15 mins every day.*

## BOP5: Caffeine – Good for your brain in many ways, in moderation

It is estimated that more than 50% of the world's adult population consume caffeine on a daily basis. That's despite us all having heard at some time or other that coffee is bad for our health. Rumours have also circulated to suggest that coffee does not actually wake us up. So what's the story? Recent scientific evidence indicates that if you regularly drink tea or coffee then the active ingredients do indeed wake you up, but only to the levels that people who have never touched a drop in their lives enjoy every single day! But don't let this put you off – there are other brain benefits that have only recently been unveiled.

*BOP5: Moderate consumption of caffeinated beverages is actually good for your brain.*

# Old Dogs, New Tricks

## Rewiring needed

New skills don't come easy when first attempted but, with a bit of dedication, the early signs of improvements soon become apparent as we start to get into the swing of things. Eventually, what once felt completely alien becomes as easy as a walk in the park.

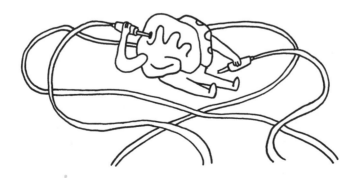

Why? Because your brain has invested sufficient resources in rewiring the pathways involved in executing that task. The key is to not lose faith when the early improvements start to taper off. Instead, you must keep pressing on. By doing so – and continuing to challenge your brain – it will continue to invest resources in improving communication between brain areas involved in whatever skill you are practising. This, for most of us, is easier said than done. As kids we were constantly confronted with having to try new things on a daily basis and so struggling with new challenges was a normal everyday experience. As adults, however, we are drawn by a natural instinct to seek comfort in behaviours that we know we are good at. As a result, we become less inclined to try new things that have the potential to make us feel like a failure. However, those who do trust their brains to adapt to any regularly encountered challenge and embrace the opportunity to try new things will inevitably continue to expand and develop their abilities.

As a child you were told by certain people who were most influential to you – parents, older relatives, teachers and peers – that you were good at some things and not so good at others. The themes that you regularly heard not only shaped your beliefs but also profoundly influenced the environments and tasks you chose to dedicate time to – the ones you became best adapted to – and the ones you tried to avoid.

If a teacher led you to believe that you were hopeless at maths then a self-fulfilling prophecy would be born. You would never again greet the prospect of having to do maths with any great relish. This shortfall in enthusiasm would result in you not trying; and a lack of effort would mean that your brain wouldn't be stretched and, as a result, unable to adapt. The inevitable poor results that followed would merely confirm your falsely held belief that you "cannot do" maths.

The exact opposite also applies, only this time in a self-propelling, upwards direction. If you really believed – because a figure of authority convinced you of your "inherent" talent – that you were good at maths, then of course you would be inspired to do more maths. The consequence of your newfound dedication being that your brain would be continuously challenged and, as a direct result, be forced to adapt to do it faster, better and more efficiently each time. With your brain now having heavily invested in some much needed rewiring, your maths skills would improve and, in turn, you'd be motivated to do more maths.

*You can't teach an old dog new tricks, huh? Well, that may or may not be true. But it's totally irrelevant to us humans. We most definitely can learn new tricks throughout life.*

It is true that the brain is particularly adaptable during childhood and adolescence. Kids seem to absorb information like sponges, which can make us adults feel as though we have permanently lost our natural capacity to learn new tricks. This is completely untrue. We just don't acquire new skills as rapidly as younger brains do, but this is largely a function of how much time we spend each day trying to pick up new skills. Children do it all the time, whereas we grownups only do so much, and far less often. Herein lies a crucial difference that explains why, through practice, kids pick things up more quickly than adults – your brain learns to learn during childhood and gets better and better the more learning it does. Unfortunately, most adult brains have fallen out of the habit of learning because of the lack of demand placed on them to attempt new challenges every single day.

*Challenge your brain regularly to learn and it will re-learn the mode that makes learning happen rapidly. That's all there is to it.*

# Brain sharpening work

Before the days of automated production lines, piece workers in pencil factories who had the monotonous task of bundling up pencils and packing them into boxes, used to struggle at first to earn a half decent wage. The reason for this was that as "rookie bundlers," being paid by the box, their productivity rate to start with was painfully slow.

The job involved them having to dip one hand into a huge container and depending on required bundle size, pull out an exact number of pencils.

New starters would have to count the number of pencils in their hand each time, whereas experienced pencil packers could dip their hand in and instantly pull out an exact quantity. Having done it day in, day out, for long periods of time, their rewired brains had learned exactly what any given number of pencils felt like.

Seeing the astonishing packing speeds that could be achieved by seasoned packers and driven by a desperate need to earn good money, novice bundler's brains were "inspired" to adapt quickly to meet the demands of this dull, but potentially financially rewarding skill. Unsurprisingly, thanks to neuroplasticity, productivity rates soon shot up.

The human brain retains its ability to learn new skills well into the latter stages of adulthood. Think about the number of people who, late on in life, discovered the art of text messaging and, in a relatively short period of time, have become fairly proficient at something that for most of their lives didn't even exist. They may never be as quick at it as young "tech native" texters, most of whom appear to have been born texting, but it is amazing to think how their older brains have restructured by creating new pathways in order to embrace something that not so long ago was completely alien to them. Especially when it involves having to master touch screen technology!

## A touch screen nightmare

*Soon after buying my first ever iPhone I seriously began to wonder if I'd made the right choice. I loved the phone but as soon as my fingers went anywhere near its touch screen, they inexplicably felt enormous! In comparison to using my old phone, sending texts and emails took ages.*

*It is only now since writing this chapter with Jack that I've suddenly realized that I'd totally forgotten about this problem. I'm back up to my old speed and my once super-sized fingers have returned to normal.*

**– Adrian**

Talking about new technology, consider a time when a friend or relative from an older generation was introduced to the internet for the first time. They might well be pretty useless to begin with, but before you know it they're forwarding you supposedly "funny" email circulars that you probably remember not being terribly amusing over a decade ago! Older people *can* learn new tricks. There's no doubt about that. All that happens is a slight reduction in the speed at which we learn and the difference in speed depends largely on how motivated a person is to embrace the discomfort of trying new things throughout adulthood.

Adults who forever enjoy developing themselves and furthering their talents – whether it be learning another language, doing some spare time studies, taking up a sport or pursuing a newfound interest – usually pick up most skills relatively quickly. That is in comparison to those who since hitting adulthood have never stretched their brains but have instead remained happily cocooned, feet up and slippers on, in their same old comfy routines.

*Understanding neuroplasticity inspires the dedication needed to change your brain to improve your skills.*

## The power of imagination

Numerous experiments have been carried out over the years with people practising new skills, varying from shooting basketballs to playing the piano. What's come to light is that whether or not someone actually physically practises a skill or instead they vividly picture doing it – after only a few days, marked changes occur in the brain. Incredibly, changes in those who had only imagined practising were almost as significant as those who had practised for real.

That the brain learns to learn throughout childhood is, admittedly, an odd concept to get your head around. As well as having to learn the very basics such as walking and talking in our first couple of years there was just so much to learn before we even got anywhere near stepping through the school gates. At which point intense learning becomes the "normal" state of affairs as we are regularly confronted by and tested with situations that take us out of our comfort zone. Not just in lessons, but outside of formal learning times as we navigate the twists and turns of learning how to get on in all other aspects of life. This might be when practising sports skills, trying to work out the unfathomable rules of attraction or negotiating our way through daily social exchanges with family, friends, strangers and those regarded as arch enemies.

Feeling out of your depth at that stage in life might not be particularly pleasant but it is nonetheless a familiar, if not a daily, occurrence. It's relatively easy to deal with when everybody else is in the same boat.

In adulthood, however, we have more freedom of choice and can exert much more free will over what we will and will not spend our time doing. Not surprisingly this means that we tend to gravitate towards activities that we are good at, enjoy doing or at least remain squarely within our comfort zone; things that through repetition have become well known to our brains – "set pieces." Unfortunately, this means we become more and more unfamiliar with the feelings of struggling to grasp a new idea or skill and so we shy away from such activities as much as we can. It's human nature to be drawn towards activities that increase our sense of wellbeing, and to be repelled from those that decrease it. Alas, in doing this we inevitably turn ourselves into set piece specialists. In fact, most people spend their entire adult lives doing things they have done many times before because these are the things that can be done with the least amount of cognitive effort.

## Sign here

See if you can do something that you've done many times before, probably without thinking about it, but this time doing it differently. Try signing your name really slowly.

Difficult isn't it! That's a set piece right there. An automatic action made inaccurate by too much thought.

We like to operate on autopilot because it's less hassle, less stressful and tends to reduce anxiety. The problem is that the easiest route in the short term is rarely the best route in the long run. And if there's one weakness that we humans often fall foul of, it's our tendency to choose the immediate, easy rewards and worry about the long term later.

## Bring it on!

Throughout most of human history, becoming a set piece specialist was absolutely fine. You spent childhood learning the basics, adolescence becoming a cog in some machine or other, and adulthood winding that cog via a repertoire of set pieces in a job for life. Food on the table to feed hungry mouths who themselves would in turn go through more or less the same cognitive transitions.

The world, however, has changed since those days. In fact the world has always been changing but, as we humans have become more and more adept at controlling and manipulating our environments, the rate of change has steadily accelerated, particularly since the Industrial Revolution when machines started getting involved. The changes used to be only really noticeable from generation to generation. Then, during the 20th century – when great technological leaps impacting on daily life occurred more regularly – they became readily apparent from one decade to the next. Now, in the 21st century, everything has gone and sped up yet again! New innovations are continually impacting on the way we work, how we socialize, how we raise children, what we do for entertainment and how we think – fast, continuous, unprecedented change that is influencing every aspect of our lives – increasing the pressure on our brains to adapt and keep up with the breakneck pace.

But, fear not, your brain is more than up for all this. If there's one thing above all others that is *most* impressive about your brain, it's the degree to which it can change to adapt to all the new challenges that will inevitably crop up. This is exactly the feature that made us humans the most dominant species on the planet. We are incredibly adaptable. Our brains will adapt to serve us in any given environment. Our collective ingenuity has led to the development of a variety of tools with which we can adapt our environments. By creating new environments and choosing what environments we wish to immerse ourselves in, we can in fact change our own brains. It may sound miraculous. It is!

## Rising IQs

IQ stands for Intelligence Quotient and is the world's most popular intelligence test. Up until the end of the 20th century it was always thought that once a person hit adulthood their IQ score would remain stable. Then a Kiwi psychologist by

*(Continued)*

the name of James Flynn noticed something curious. When he compared IQ scores of the same people from one IQ test to the next – they had gone up. We now know that IQs around the world have increased by an average of three IQ points every ten years!

But what is driving this increase in intelligence? As far as we can tell it is driven by our use of technology. From decade to decade the amount of information we have access to has dramatically increased, firstly via television and then through the internet. The more information our brains have to juggle on a daily basis, it seems, the smarter we become.

## Finding Flow

To get the best performance out of your brain, you should bear in mind what modern neuroscience has taught us are the rules of the game. Your brain will physically rewire the connections between brain areas involved in any mental ability so that the interactions between them are:

- faster

- stronger

- more efficient

So long as you take care to perform the desired mental function:

- regularly (ideally every day)

- intensively (stick at it for a decent amount of time, don't give up when it gets tough)

- over long periods of time (keep up the regular intensive training for weeks/months)

This essentially describes what is popularly described these days as "brain training." It might sound terribly impressive and it might even have misled you into thinking that you need some special kit to do it properly.

## Working memory

Working memory is what you use to keep a phone number in your head for long enough to dial the number. You also use it any time you plan what you are going to do with your day or try to solve a problem. Sights, sounds and other pieces of information can be held in mind (on your Hold Line) for just long enough to perform a mental task effectively. Experiments investigating brain training have found that exercises that increase the capacity of working memory can improve performance in a wide range of cognitive tasks. So much so that, if we improve working memory, our IQ score is likely to increase as a direct result.

With a whole bunch of useful brain teasers in an easily portable package, brain training devices do make training your brain on your way to and from work more convenient. They can certainly improve your devotion to self-improvement and, yes, leading companies may well have several hundreds of millions of subscribers for a web-based suite of games. But that doesn't mean you are not training your brain whenever you engage in anything that taxes your brain regularly, intensively and over long periods of time.

As you improve at any skill, hobby or mental ability you are engaging in brain training. There is no magic in the commercially available games – just convenience. The convenience might make the difference between training hard and long enough to make a tangible difference in how your brain functions, but if you are determined enough you can do it with or without these products.

The key to doing enough mental work to actually make a physical difference to how your brain functions, is finding "Flow." Flow is a psychological term coined by Mihaly Csikszentmihalyi (pronounced "six-sense-mihal-i") with a distinct meaning that has been bandied around so much it has lost much of its original simplicity.

 The average capacity of working memory is seven items. By regularly trying to hold more items of information in memory at the same time you can increase your working memory capacity – gradually leading to improvements in your ability to solve problems.

To remember what Flow really is all about, just remember Goldilocks. She wanted her porridge to be "not too hot, not too cold, but just right." Flow comes when you challenge your brain with some kind of mental task that is "not too hard, not too easy, but just right" – keeping you keen to keep on training. If it's too easy you may well be able to keep your head down for a solid half an hour, but your brain will only invest resources in changing connectivity if it is pushed out of its comfort zone. However, if the challenge is too hard and it makes you feel stupid, convinces you that you are getting nowhere and that the task is futile, then you'll end up all frustrated, wound up and looking for something less productive to do instead. This is something that the new commercial brain training games are very good at – computer game designers are masters at finding that line where the gaming is challenging without leading to despair.

Whatever brain training you choose to do, whether it's getting to grips with a new software package, memorizing a speech or perhaps learning how to cook a particular dish, it will only change your brain if you find Flow. This is because Flow keeps you at it for long enough to trigger brain changes. Having found Flow, you will no doubt have experienced considerable enjoyment from the process and taken away some real satisfaction from your achievement; it is highly likely that you'll soon be up for another challenge. Finding Flow in the next challenge will be key to ensuring that your self-propelling belief keeps you going in the right direction and that your brain continues to invest in whatever rewiring is required.

## You can't do that!

*At the age of 43, my first book Polar Bear Pirates and Their Quest to Reach Fat City was published. Despite the fact that I had been told on more than one occasion that I would never be capable of writing a book, it became an international bestseller.*

**– Adrian**

## Chapter takeaways

- By improving working memory – as far as your IQ is concerned – the only way is up.

- If you wish to progress you'll need to learn to learn again and feel comfortable being out of your comfort zone, just like you did as a young child.

- To discover the positive effects of neuroplasticity and continue to improve at something, you need to do it regularly, intensively and over long periods. That's "brain training."

- Step beyond set pieces. Stretch yourself – the easiest route in the short term is rarely the best route in the long term.

- Remember "Flow" – set yourself goals that are not too hard, not too easy, but just right!

- Yes, you can teach an old human new tricks.

# Cyber Heads

## Brain for sale – no longer needed

Is technology good or bad for our brains? With around 2.5 billion internet users, fifteen million texts being sent every minute and many people now spending more time online gaming than they do sleeping, are we all rapidly becoming mindless zombies unable to interact with others on a face-to-face basis? Or, are people getting all steamed up about the potential consequences of digital immersion that in reality pose no major threat whatsoever to the future of the human race?

 The more Facebook friends a person has, the greater the grey matter density in brain areas involved in social interactions.

Internet search engines make a whole world of information instantly available to us, information that is literally at our fingertips. So why would anyone want to bother committing anything to memory when it can be pulled up on a screen within nanoseconds? Labour-saving devices have unquestionably changed our lives beyond all recognition, but will the continuing tsunami of innovations leave us all with redundant brains that are unable to do anything unaided?

As yet there are no conclusive answers to these questions, but whether or not all this technology proves to be good or bad for brains, you can rest assured that your brain will have been doing what brains do so brilliantly well – it will have been changing and adapting to meet the demands of the new technological environment it is now operating in. No matter how old your brain happens to be, it will already have been busy reconfiguring and shaping up to embrace whatever new challenges this ever-expanding techno era happens to throw at it.

# I want it now!

Having said that, our brains might have changed over the years but human nature hasn't. The main reason why we, out of approximately 8.7 million other species currently sharing this planet, have been so successful is largely down to a deep-rooted desire to progress. Thanks to lightning speed technology, we are now all advancing at an ever-increasing rate with everybody expecting everything to be done in an instant – we want it now! And, if we don't get what we want quick enough, human nature dictates that we will look for a shortcut, and as soon as one becomes available, we'll take it!

With shortcuts at the very heart of human nature we are all unsurprisingly more than happy to take the fastest option that modern

## Fast food expectations

When you read a microwave meal's instructions and it says "Microwave for two and a half minutes, stir well, re-cover and then microwave for a further two and a half minutes before stirring again and serving." Are you disappointed that's it's going to be such a laborious process?

Are you dismayed that it's going to take so long, that you're going to have to wait at least five and a half minutes before getting to eat it, and that you're actually going to have to bother taking it out half way through to stir it?!

technology can provide. And, with the prospect of life improvement being at the forefront of our minds, most of us will, at the earliest opportunity, interface with whatever the latest technology happens to be, in the belief that it's going to make life easier or more interesting.

We may always be looking for the fastest, easiest route forward and by using a whole array of "external brain" devices at our disposal, ranging from PCs to phones, we may have found numerous new ways of shifting a lot of the workload to them. But that doesn't necessarily mean our brains are going to have less to do, nor does it mean that they are going to become lazier and inevitably, having become semi-dormant, find themselves out of a job.

On the contrary, a lot of the groundwork might be done for us, but by shifting it we simply free up cognitive resources for a new pile of work along with a whole new set of pressures and obstacles created by it. All of which will need to be overcome and

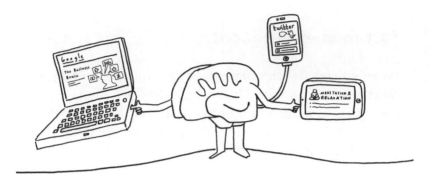

dealt with at an ever-quickening rate. When calculators first came onto the scene, there were serious concerns that they would make brains lazy and more recently, when internet search engines first appeared, there was much talk about them making minds intellectually stagnant. In both instances, these worries have been unfounded, the reality being that new doors have opened and brains being brains have moved on to bigger, more exciting challenges.

Despite how much technology comes onto the scene, if we fully utilize it, our brains should be moving in the exact opposite direction of idle. Having once been stretched to their limits to perform a particular function, they should find themselves having to rise to completely different tests in fresh areas.

For example, where a brain was once challenged by the prospect of map reading, advancements in technology will mean that despite no longer having to be quite so proficient in that skill, it will have to rise to the challenges of operating and efficiently following the instructions of a satnav. Those of us who have experienced setting off from A to go to B – only to end up in C – will know that this is no mean feat. Despite thousands of years' worth of inherited instinct and our gut feelings screaming at us that we are heading in completely the wrong direction, we still keep faith with the technology!

## Forever lost

Your satnav doesn't have common sense – you do. Map reading and navigation are useful skills to have, particularly in the event of a technical hiccup. If you do want to hold onto them, maintaining your self-navigational skills is simple. Don't rely on satnav all the time, especially when you want to get to places you've been to many times before. Think of all those drivers of London's black cabs whose inflated hippocampi shrink back down after they retire – keep on refreshing those navigation pathways of your brain. Before switching on the satnav, take a look at your route on a map to give yourself an idea of where you're going and, hopefully, the next time you find yourself driving the wrong way down a one-way street, in the middle of a building site or along a road that doesn't exist, common sense will prevail.

The bottom line is that whether or not a brain does get made redundant is of course up to its owner. Neither technology nor the brain itself can be blamed if through lack of activity it does get left behind. Provided you give it the opportunity to be stretched, it is more than capable of keeping up.

## External brain reliance

As far as brain health is concerned, making use of technology is not in itself a problem. What is currently causing some alarm in some circles is the increasing number of people who are becoming permanently hooked into, and addictively dependent on, technology.

Even the less observant will have noticed the blind reliance that more and more people are placing in the hands of these devices. Taking a few moments to notice and reflect upon the behaviours of people within your immediate vicinity will surely convince you that an obsession with gadgetry is taking over people's lives. Walk down any busy street in any town and it won't be long before you see someone scurrying along the pavement, head down, squinting at some device or other – only to step out into oncoming traffic without looking.

With minds elsewhere, these digital lemmings seem completely oblivious at times as to just how close they are to eradicating themselves from the human gene pool. Perhaps their need to be permanently technologically engaged drowns out their awareness of everything else around them, or maybe an overestimation of their own abilities to multitask is leading to more subtle cognitive drawbacks than being bounced off a bus.

Whatever the reason, many people are becoming too dependent on new technology for their own good. One of the classic measures of overdependence is automated, unthinking behaviour. How often have you seen people in your midst failing to resist the temptation to pull out their phone the moment they hear it beep, buzz, ring or feel it vibrate? Have you noticed that this happens irrespective of whether the circumstances make it appropriate to do so? The most popular times seem to be during meetings, in restaurants or whilst attempting tricky driving manoeuvres.

Should such unsociable, at times rude, and potentially dangerous tech habits be tolerated on the basis that "you can't stop progress?" There is no definitive answer to this as it is down to each and every one of us to decide for ourselves. There is, however, one thing that is becoming apparent, whether it is you, family members, friends, colleagues or random strangers you've witnessed doing this – it's a fairly safe bet to make that you are now so familiar with these scenarios that you're beginning to accept them as the norm.

## Yesterday's "black art" – today's norm

*As an IT sales person back in the 1980s I earned a lot of commission selling fax machines. Having witnessed my well-rehearsed product demonstration involving the transferral of a document between two "facsimile" devices, people's eyes would light up in disbelief as they struggled to comprehend what they had just witnessed. As if by magic, the very same document sent from one machine would slowly emerge from the other. The endless possibilities for those beholding this wonderment would immediately become apparent, and I could once again look forward to smashing my target. Alas, the window of opportunity, as with all the other IT products I went on to sell, would suddenly close. What was once regarded as the very latest "black art" technology soon became the norm.*

*It's hard to comprehend that a fax machine is something that people once got very excited about, that not that long ago people were astonished to find themselves talking to one another whilst "out and about" on phones with no wires, or that it was once amazing to see and hear someone right across the other side of the world on a laptop screen.*

*IT product development is now moving on so fast that whatever you see on sale in the shops or online has already been superseded, you're looking at yesterday's technology. The most interesting thing about this once fascinating fact is that with your brain's ability to embrace all things new, it probably comes to you as no surprise whatsoever!*

**– Adrian**

# Will technology ruin our brains?

To date, most of the available evidence on whether digital technology is good or bad for your brain is purely anecdotal. Studies are being conducted at this very moment to provide hard data that will establish whether our obsessive use of gadgets is having unintended consequences. In the meantime, there are a few studies that have already hit the academic press from which we can begin to forecast future findings.

Technology in itself is neither good nor bad, the problem lies in how we use it. Your malleable brain, as you are now more than aware, will accommodate the demands of any environment, this being the case whether the environment in question is physical or virtual. This ongoing accommodation will happen for better or for worse, whether you like it or not, as long as you continue to engage regularly, intensively and consistently with any given technology.

In addition to this, as we know, old habits – once formed through repetition – die hard. For instance, eating habits adopted in childhood (when metabolism is relatively high) almost always continue into adulthood (when metabolism inevitably slows down). The consequent excess of calories leads to an ever-expanding waistline, a scenario familiar to all but the most disciplined of eaters. The same principle can be applied to technology. Once a person develops a reliance on technology, not to mention an expectation of regular messages and online updates, they can end up panicked by any interruption to the flow of communication. They may be thrown into a hissy fit when unable to get a connection or, worse still, fall into a spiral of depression when a whole day passes by without hearing the reassuring ping of messages arriving in their inboxes.

## The elephant in the room

**Access:** In 1984, there were one thousand devices hooked up to the internet across the globe. Eight years later in 1992, this number hit the one million mark. The one billion mark was crossed in 2008.

**Quantity:** There were 4 exabytes of new data created in 2012 – that's four billion, billion bytes – more new information created in a single year than in the 5000 preceding years put together.

# Staying in to play

With some people spending up to two months a year glued to them, the big concern for a long time was the amount of time people spent watching television. With the average household now having more screens in it than people, the latest worry is over the amount of time being devoted to gaming. With so many spending huge chunks of their lives participating, a major concern with video gaming in recent decades has been that the violent nature of many titles might be leading to a new generation of morally corrupt individuals. It turns out there is little evidence to support this.

In both cases of excessive TV watching and video game playing, the real root cause of problems revolves around displacement. Displacement of time that could be spent socializing face-to-face, for instance, and thereby gaining valuable experience enabling the all-important "soft skills" of communication to be developed. Major problems with technology arise when digital immersion displaces *all* the spare time that might otherwise be devoted to real-world engagement such as engaging in group activities or taking part in sports. The brain pathways involved in such activities either don't develop properly in the first place or start to fade away.

It's very much about getting the balance right. There are in fact several benefits to be had when gaming enthusiasts clock up many hours playing *action* video games. It forces their brains to adapt to the perceptual and cognitive demands of such virtual worlds, leading to unexpected, positive enhancements in several areas. In contrast to *non-action* video gaming, superior visual perception, visual short-term memory, spatial cognition, mental rotation, multitasking and rapid decision-making improvements have all been demonstrated to result from intensive *action* video gaming.

## The trouble with the youth of today

*When driving in my car I often stop just down the road from where I live to let a group of teenagers coming home from school cross the road. Every time I do this, I get the distinct feeling that both my car, and myself, are invisible. I don't think in the past two years any of them have ever put their hand up in acknowledgement of me stopping for them or even given me so much as a nod of recognition that I exist. This worries me.*

*It could be that they simply lack the confidence to engage with others outside their group. Maybe they just don't have the social skills to do it or perhaps common courtesy isn't particularly high up on their agenda. They appear to be so wrapped up in their own worlds that they are oblivious to everything around them. I've often thought that the possible source of this disconnected, insular behaviour could be today's technology. It would be easy for me to attribute it to them being tech natives and assume their preferred mode of communication must be via email or text.*

*Yes, I could put it all down to technology, but then I'm always reminded of this:*

> *"Our youths love luxury. They have bad manners, contempt for authority – they show disrespect for their elders and love to chatter in place of exercise. Children are now tyrants, not the servants of their households. They no longer rise when their elders enter the room. They contradict their parents, chatter before company, gobble up food, and tyrannize teachers."*

*It's a quote attributed to Socrates the famous Greek philosopher who lived from 469BC to 399BC. It would seem that in two and a half thousand years, technology or no technology, some things just haven't changed!*

**– Adrian**

In addition, the concerns of parents worrying that too much time being spent staring at the screen might ruin their kids' eyesight was also proved to be completely unfounded. On the contrary, it may actually improve certain aspects of vision, so much so that video gaming may now be prescribed as therapy to help people improve visual problems associated with conditions such as amblyopia ("Lazy Eye").

Another aspect of displacement worth mentioning is that unless you're managing to exercise at the same time, spending hours on end in front of a screen, whether it be a TV, PC, laptop, tablet, games console or phone screen, it isn't going to do anyone any favours in the physical health department. Excessive amounts of screen watching has been directly linked to obesity, increased cardiovascular disease and Type 2 diabetes.

## Fatal attraction

In South Korea, a 41-year-old man and a 27-year-old woman became so obsessed with an online role-play game involving caring for a virtual girl that, in a horrifyingly ironic twist of fate, they accidentally starved their own, real-life, three-month-old baby girl to death.

In 2005, another South Korean called Mr Lee dropped dead of heart failure after playing a game called Starcraft for over 50 hours straight.

In 2007, a Chinese man called Zhang died suddenly after playing World of Warcraft continuously for seven days.

These cases, and many others, have led to the term "Internet Addiction" finding its way into the psychiatric profession's official diagnostic manual (DSM-5).

You'd be forgiven for thinking that this is a problem specific to East Asia, but this isn't the case. Digital innovations are made available and embraced earliest in these countries, before much of the rest of the world gets their hands on them. Observing the negative outcomes of digital immersion amongst early up-takers gives the rest of the world a valuable heads up that might help others take measures to avoid such lethal scenarios.

## Multitasking?

Not that long ago it was rare for people to watch TV whilst simultaneously surfing the internet; now such dual screening is commonplace. Studies investigating this behaviour have revealed that during a 30-minute viewing period people will switch between the two screens on average around one hundred and twenty times. That's about four times per minute or once every fifteen seconds! Might such behaviour herald a new era of prolific multitasking that enables us to squeeze yet greater efficiency out of our busy, information overloaded lives? Probably not.

The first thing to get to grips with is that there is no such thing as true multitasking. Our brains have yet to evolve the capacity to actually perform two completely different cognitive processes simultaneously. Mental tasks that feel like they are being done in parallel actually involve rapid switching between the two. And, anytime a human brain switches from one task to another, there is an associated cost. You don't quite pick up where you left off when your mind returns to the "other" task. There is always a slight delay in remembering exactly where you were in the thought process and in recommencing that cognitive process.

Women are famously good at multitasking behaviours, but the fact that these are all happening in parallel is purely an illusion. The reason women are good at doing multiple things simultaneously revolves around a superior ability compared with most men to minimize the cost associated with each switch. Whilst some people, through regular, intense and consistent training have become extremely adept at making this process as efficient as possible, there is nonetheless always a small, unavoidable, cognitive cost associated with each switch between tasks. In men the switch costs tends to be larger than in women, but in both sexes there is a measurable negative consequence of switching between multiple cognitive tasks rather than focusing on one task through to completion.

## The buzz and bleep of modern living

Smart phones are a constant source of unintended distraction. "Unintended" is the operative word here because "intended" distraction is ideal for encouraging certain useful brain states, as we'll discover in the following chapter. Creative thinking really benefits from a bit of distraction, especially when a person's brain is cluttered with anxious thoughts and stuck in rigid ruts, struggling with seemingly unsolvable problems. The "intended" distraction shifts the brain into a different gear, de-focuses the cognitive machinery to let unrelated ideas flow more freely.

The problem with "unintended" distractions from phone alerts notifying you of a text message, email, call, or online social networking update is that your attention is repeatedly pulled away from your thought processes. Each time this happens your brain incurs a switch cost. If you allow your environment to constantly interrupt your thought processes then all those little distractions will add up to a very unproductive day at the office.

But won't our ever-changing brains adapt to help us perform better in any environment? Surely brain pathways will be reconfigured to help us block out these minor distractions? It is possible, but unless you adopt a specific strategy to train up such useful brain adaptations over many weeks and months, all the research carried out so far indicates that this is highly unlikely to happen spontaneously.

## Instant response pressure

The virtual world of social and professional networking is cunningly designed to lure us into patterns of use that make us addicted to constant interaction. It is generating a whole new, previously unknown world of stress. Teenagers and adults alike are often expected to reply immediately to any text, email, instant message or online social networking message. Teenagers who do not respond straight away risk being socially excluded and adults face the possibility of losing out at work or in business. Many people now feel a tremendous pressure to be available constantly, day and night, and throughout the weekend.

The ultimate cause is other people's expectations of an immediate response. It would be wise for all of us to do whatever we can to change other people's expectations and
*(Continued)*

create some distance between us and the cyber-onslaught of constant connectivity. What creates these expectations in the first place is the all-powerful influence of instant gratification. Activity increases along the brain's pleasure pathways (Reward Line - please see brain tube map on page 8) as they are temporarily satisfied by a quick message response. Conversely, not receiving an expected reply, leads to decreased activity along the Reward Line, causing feelings of disappointment and anxiety.

If we are to have any hope of changing other people's expectations we need to make it clear to all – "Sorry, but you probably won't receive a response from me straight away, so please don't expect one. I will however get back to you within the next 24 hours. Thank you."

We can't change a culture of high expectations overnight but, little by little, starting with friends and family then moving on to professional connections, once everyone realizes just how important this is we could eventually regain the right to reply – at a time that is right for us.

## Brain training ourselves to distraction

Scientific investigations into the effects of multitasking and constant interruption on our cognitive abilities is in its infancy. Early findings are extremely interesting and provide some valuable clues about what we are likely to discover as the experimental data continues to accumulate. In a comparison of heavy-to-low media multitaskers it was found that people who regularly use technology to do multiple tasks at the same time are less able to ignore distractions than those who don't. The task in question involved a display of lines tilted at different angles and surrounded by a varying number of additional lines to serve as distractors. The performance of the heavy-media multitaskers declined when more visual distractors were added. However, the performance of the low-media multitaskers remained stable no matter how many additional distractors were added. In other words the low-media multitaskers had retained the ability to stop the distractors from interfering with the cognitive task, but the heavy-media multitaskers had lost this ability.

One likely explanation to account for this data is that the daily, intensive, consistent media multitasking behaviours has led to brain changes that make them more, not less, sensitive to distraction. They may have unwittingly trained their brains to refocus automatically on any external information that arises in their environment. The only way to prove that these behaviours actually *cause* increased sensitivity to distraction would be to compare measurements before and after people started engaging in heavy multitasking. You can be certain that someone, somewhere, will be conducting research into it right now. In the meantime you might want to think twice about checking your emails a hundred times a day.

 Heavy media multitaskers are generally worse at controlling their impulses and score lower in tests of fluid intelligence than light media multitaskers.

Assuming that further studies investigating these phenomena all point in the same direction, the implications are clear. We need to stop and think about how we use the wonderful tools of digital technology so that we harness all the many benefits without falling into the trap of inadvertently training up cognitive processes that overall serve us poorly. We can all then establish our own set of rules to help us maximize the benefits and minimize the cognitive drawbacks. For instance, next time you get a new phone – actually read the instruction manual. Not the whole thing, just the part that guides you through the process of changing which alerts can make it buzz and bleep, and which silently register the message without causing an "unintentional" distraction. Switch off all alerts that are rarely time critical.

Ask yourself – do your text messages really require instant attention? Do you have to be alerted to every email that lands in your inbox the second it arrives? Is the news that a random stranger is now following you on Twitter urgent enough to risk halting a fascinating conversation, to steal precious time spent with family and friends, to derail a productive flow of thought, crash a car or walk out into the road? Or can these relatively low-priority interruptions wait till later?

## Digital addiction litmus test

How often do you feel you need to pull your phone out and check it for messages?

Having given yourself an honest answer, try this: switch your phone to silent and hide it away somewhere for sixty minutes at a time for just one day. Only allow yourself to have a look once an hour – no peeking!

Mankind has managed to get by for thousands of years without handheld communication devices and yet research suggests that on average we check our smartphones every six and a half minutes!

Over the next five days keep a tally of how often you do find yourself picking up your phone and checking it. You might surprise yourself as to just how addicted you've become. Surely once an hour is enough to stay on top of everything. Isn't it?!

## Chapter takeaways

- Your brain will already have been busy physically changing to make everything in its regular environment "the norm."

- By shifting your workload to "external brains," the pace will quicken and new challenges will emerge – but beware of becoming over-reliant.

- Don't go blaming technology for the behaviour of teenagers. Remember – they've been around a lot longer than technology has!

- Brains don't do multitasking, they do switching at a cognitive cost.

- Are you the master or the slave of your digital devices? Make sure you're the one calling the shots or else your brain will adapt in ways that serve you badly – don't lose your focus.

- Unintended distractions add up. What's the sum total of yours so far today? How many A-ha! moments were missed via mid-thought disturbances – put it on silent!

# Gone Fishing
# (for Great Ideas)

## Dunking

When trying to come up with new ideas, there's something you can try out that we call "dunking." As well as being wonderfully relaxing, dunking can be a very effective way of being more innovative. It's a technique inspired by the brilliant engineer Thomas Edison whose inventions truly changed the world.

He would sit in a comfy chair with his arms hanging over the sides whilst clutching two large ball bearings, one in each hand, strategically placed, dangling over a wooden floor. He would start dozing off, but as soon as he actually fell asleep his hands would release their contents and the noise of the ball bearings clattering on the hard floor would wake him up with a jolt. By dunking in and out of sleep like this, he found he could seize any inspirational nuggets that popped into his head whilst in this semi-sleep state.

Just like dunking biscuits into a hot cuppa, the secret is not to stay dunked too long otherwise you drop off completely into deep sleep and lose any ideas you may have had. You might well have already noticed that your best ideas come to you when you're daydreaming in the bath, on the loo or commuting. These scenarios provide great opportunities for cracking difficult problems because, in each instance, you might just doze a little without completely drifting off. The reason being, these are places where you remain conscious of the potential dangers around you, not to mention the possible embarrassment of nodding off completely.

The other daily opportunity to do some dunking in the privacy of your own home is during that most precious "snooze time" in the mornings when the alarm has gone off. You're half awake, still in a lethargic state and desperately trying to make the very most of those all too brief moments before it goes off again. Try and use this time for a quick dunk by pointing your relaxed, wandering mind in the direction of some of the stuff you've been working on the night before and you will hopefully capture some inspirational ideas.

## Sit on the edge or dive right in?

There are actually two slightly different dunking techniques. Either way you need to get yourself nice and relaxed until you find yourself in a "twilight place." Having given a subject considerable thought, you can either imagine yourself sitting on the edge of your subconscious waiting for ideas to float by and quickly netting them before they disappear off into the depths of your mind. Or, if like many, you are a firm believer that solutions to problems already exist – "you just have to go and find them" – try doing what most seasoned dunkers prefer doing: imagine diving down into your subconscious, have a good look around and go grab that pearl!

This may all sound a bit bonkers but, despite what anyone may think, it really does work and the more you do it the easier it seems to be to do, and the more ideas you seem to find.

## Why is dunking so effective?

Although right now you will be completely oblivious to it, your brain is constantly working away beneath the surface of aware-ness, posing questions, solving problems, activating memories and freewheeling thoughts. Dunking enables you to briefly dip below the surface and capitalize on all this unseen hard work.

"Consciousness" is all the stuff you are aware of: the sights you see, the sounds you hear, the thoughts you think and that voice inside your head that narrates the moment-to-moment events of your day. The "subconscious" is all of the work that your brain does which you cannot, under normal circumstances, be aware of.

The reason that so much of what our brain does goes under the radar, is that consciousness has a limited capacity whilst the subconscious has a much, much, much larger capacity. If conscious awareness was cluttered with every last little detail that your brain processes in any given second, your mind would be in a permanent state of chaos. Conscious thought is spared from being drowned in information to ensure it has sufficient resources available to deal properly with only the most pressing matters.

Conscious thought is just the tip of the iceberg in terms of what your brain is up to every moment of every day, and a sneaky peak beneath the surface can be extremely rewarding.

Even when you are feeling totally uninspired your brain is nonetheless always busy dealing with a constant, never-ending onslaught of sights, sounds, smells, bodily sensations, memories, emotions, plans, ideas about what to do, where to go and whom to speak with, all constantly processed by dedicated brain circuits. Your subconscious, believe it or not, is always at it – forever considering new ideas and trying to find solutions to your problems. Whether or not you become aware of the fruits of all this brain activity depends upon what you are doing. If your conscious thought is fully engaged: having a conversation, performing a task, looking at a website, reading a magazine or, for that matter, this book, then the ideas that quietly bubble up from the depths of your subconscious will probably end up going unnoticed amongst all the many distractions. It's like trying to get the attention of a friend you've spotted on the other side of a busy high street by whispering to them!

However, when you are in between conversations, in between tasks and in between thoughts – that's when your conscious thoughts are sufficiently calm for those subconscious idea bubbles to cause a noticeable ripple. *This* is why your greatest ideas usually come to you when you are not up to much.

*With your subconscious being so busy churning up ideas, theories and concepts that might come in useful – the trick is to quieten your brain enough to catch them.*

## Perfect timing

Sleep is vitally important when it comes to being innovative, the very best brain state for creative thought being when you're asleep. To be accurate, the most creative brain state occurs just at the moment you fall off to sleep when it enters the "hypnogogic" state. So that is why dunking is so effective – not only are you in a relaxed, quiet state all ready to receive ideas – your brain is also getting in the just right mode to create them.

Your brain doesn't go into sleep mode all in one go, different parts of your brain enter sleep mode and disconnect at different times. As brain regions each power down one by one over the course of about 20 minutes, certain areas find that they are no longer able to send and receive messages to brain networks they are usually in intense communication with. Effectively this means that parts of the brain that are not yet in sleep mode are ticking away in isolation – each doing its own thing without any of the normal coordination with other brain areas. This interim state between all of the brain being awake and all of the brain being asleep appears extremely effective in catalyzing creative thought.

Whatever the precise reason for having different parts of the brain awake and asleep being so conducive to problem solving, we can at least feel confident that the hypnogogic state will deliver the goods because it was indeed extensively exploited by Thomas Edison. He was without doubt one of the greatest inventors of all time. By the end of his career he had an incredible 1,093 patents to his name. To get there he first had to crack a particularly tricky

problem relating to exiting the hypnogogic state just before entering the deeper state of sleep beyond which all those great ideas
drift away.

The problem that Edison quickly observed during his sorties into
the creative phase of early sleep was as follows: if he nodded off
just for a few minutes he would often wake up with an innovative
idea. However, more often than not, he would wake up half an
hour or more later, by which time any fantastic idea he'd had was
forgotten. Being a brilliant inventor he inevitably dreamt up an
ingenious solution to this predicament and that solution was, as
we now call it, "dunking."

## Enhanced creativity

Edison noticed that whenever he observed people falling asleep
in public, at the very moment they "dropped off," their neck muscles
would relax causing their head to drop forward suddenly –
"nodding off" – waking them back up again. In fact, all the skeletal
muscles – those that control the arms, legs, body, neck and face
– become paralyzed when we fall asleep. This accounts for wide
open "fly catching" mouths and small but highly embarrassing
dribbles of drool emerging from them. Edison realized that this
presented a great opportunity to provide a wake-up call at exactly
the moment that any ideas are on the brink of being lost forever
due to an overly long snooze.

His solution was simple. He would first work very hard on the
problem he was trying to tackle. He would spend many hours
looking at it from all different angles, mulling over a wide variety
of bizarre and wonderful solutions to his problem. He would read
broadly on the topic and talk to anyone who might have some
useful input. When he began to feel sleepy, still with all his reading

and thoughts on the matter fresh in his mind, it was then that he used to settle down, ball bearings in hand for his pre-planned nap. From the brain's perspective, the reason this strategy was so effective was that he was woken just after he had dipped into the hypnogogic state, which happens to occur simultaneously with the brain losing control of the hand muscles, sending the metal balls crashing to the floor with a bang.

If you are dubious about how effective Edison's homemade "creativity machine" really is, bear in mind that he invented the light bulb, the phonograph and the dynamic image camera. In other words, without his prodigious dunking-enhanced creativity we would have to have waited around an awful lot longer for electric lighting, recorded music, televisions and the birth of the movie industry. If it worked for Edison, then there's no reason why it shouldn't work for you too!

If by any chance Edison's ball bearings seem a bit low tech, logistically impractical or perhaps even embarrassing, then try this out. The next time you decide to have a power nap, give a problem you'd like to solve a lot of serious thought and then, depending on how good you are at nodding off, set your mobile phone alarm to go off in 15–20 minutes time. You should, with a little practice, find yourself being able to enter your hypnagogic state and wake up just in time to find some useful ideas floating around in your head. If you get in the habit of doing this once or twice a day you will find that within a week you start popping back up from hypnagogia automatically, just before the alarm goes off. But do continue to set your alarm for peace of mind. Please note: never nap for longer than 20 minutes or you'll shut too much of your brain down and you'll have trouble waking back up again. When you do eventually wake, you'll feel more sleepy than before you nodded off. And, regrettably, any stunningly brilliant solutions will have slipped silently away.

# Where in the brain do these creative thoughts actually come from?

The most important brain area for creative thought is just behind the top right corner of your forehead – the dlPFC stop (Dorsolateral Prefrontal Cortex) on your brain's Hold Line. If you attach a bunch of electrodes to a person's head whilst they are trying to solve a tricky problem, a burst of electrical activity within the dlPFC in the right half of the brain peaks a whole second before that person actually becomes aware that they have found the solution. In other words, this appears to be a hub for all your unconscious, creative thought.

To further bolster this claim, a rather controversial study zapped the brains of a group of volunteers as they tried to solve a sequence of logic puzzles. They used a pair of electrical currents to either: simultaneously deactivate the right PFC and activate the left PFC, or *vice versa*. They found that when the activity in the left PFC was dampened and the right PFC was boosted, participants could solve the logic puzzles faster and so complete more of them. This was in comparison with those people in whom the electric currents were applied the other way round. We'd like to make it perfectly clear here that we're not recommending self-electrocution for creative thought! Although, frighteningly, such devices are being made available to the public and are marketed with the promise (of all things) that they will make you a better computer gamer.

# Using novelty to jolt you out of set piece behaviours

The left PFC is known to run tasks that a person is overfamiliar with – set piece behaviours that have been performed so many times before that they can be fully executed purely on autopilot. The right PFC, on the other hand, kicks in when a person needs to process new information, go to a new place, or carry out a novel task. To get the right PFC firing on all cylinders it is important to experience new things. Try working in a new environment, e.g. a park, meeting room, or a cafe that you have never been to before for an hour or two each week. Once in these fresh, novel environments, actively try to think of new solutions to old problems. If you can't spare the time to leave your desk, take a 5–10 minute break to surf the internet for a new online game or brain teaser website, purely to coax your right PFC into action by forcing your brain to perform a novel task or process novel stimuli. Listen to music from a different culture. The unfamiliar rhythms and voices singing in a foreign tongue may well be the unusual sensory experience that wakes up the right PFC sufficiently to help get the creative juices flowing.

How often have you found yourself in the situation where you've got stuck on a particular problem and, having reached what seems like a dead end, you're unable for the life of you to see any possible solution? Then, having walked away from it and done something completely different, you come back to it only to find the answer is right there in front of you, staring you in the face! It's a struggle to understand why you weren't able to see it in the first place, now that it seems so glaringly obvious. This is why, whenever you are trying to concentrate on something, it's best to do it in 45-minute chunks with a 15-minute break in between. For most people, working this way is so much more productive. The key being that the 15-minute break is *only* 15 minutes!

## Switch it up

The best way to start thinking outside the box is to actually physically get outside your box: in other words, get out of your office!

Some fresh air up the nostrils and a lack of walls and ceilings definitely induces some great "corkscrew thinking." It's a term that was originally coined by Winston Churchill and is still used today to describe looking around obstacles and barriers to find solutions.

Just like food tasting better when cooked outdoors, thinking also seems so much better when it's done outdoors. If you want the ultimate outdoor meeting, try and hold it beside water, especially by noisy water – for instance, by a river, waterfall, or on a beach with the sound of waves crashing in the background.

The reason that this works is because it's a fantastic way to wake up the right PFC. The critical feature of moving water is that it is random and constantly changing. This means that watching the moving water or listening to the sound it makes involves an ever-changing pattern that cannot possibly be accurately predicted from second to second, which keeps your brain's pattern detectors guessing, thus coaxing the right PFC into action.

## Brain shaking

Brainstorming may involve getting everyone together to bounce ideas around in an environment where everything is considered and no one is criticized for coming up with outlandish ideas. What works even better is a "brain shake." Everyone considers the problem in their own time, coming up with ideas individually in advance of the actual discussion. These are written down anonymously and sealed in envelopes before being opened up on the day of the discussion. This ensures that all the best ideas get to see the light of day, as opposed to classic brainstorming sessions where study after study has suggested creativity is actually stifled.

What is key for a good brainshake meeting, whether it's indoors or outdoors, is to really disturb and shake up the *status quo* – agitating and invigorating minds with something new – and to encourage the people involved to have some fun, which breaks down the mental boundaries of rigid thought. If people are given the opportunity to have a laugh, then they become physically relaxed, which in turn means their brains can get into a relaxed state; ready to receive ideas bubbling up from the depths of their subconscious brain. When people start laughing and joking in a meeting there's a fair chance that a pearl of an idea is on its way. Even cracking a joke requires quick thinking, unexpected associations and perfect timing – a highly creative process in its own right. People will laugh, but then realize: "that isn't such a daft idea!" All too often, solution-finding meetings get brought to an abrupt end because people aren't taking them seriously enough – a big mistake.

By relaxing and having some fun, the innovation centres of people's brain are given a bit of a nudge. Out-of-doors meetings, as long as they don't become the norm by always being held in the same place, really do create that much needed novelty factor to help bump-start the mind. Just like when training your body in the gym, it's good to vary your exercises, to do something different, to flex muscles not usually involved in your set routines.

## Inputting relevant data to feed eureka moments

Unfortunately, there is no such thing as a free lunch. If you think that all your problems will be solved with a quick crossword and a nice nap then you've skipped one vital step in the process.

If you want to unleash your subconscious on something you must first put the work in. It's down to you to input all the relevant data so that the novelty-processing, creative brain areas have all the materials they need to find a solution. The more you research a problem, the more you read books and magazines, scour the internet, contact experts, discuss and bounce ideas around with friends, family or anyone who will listen, the more your subconscious will get busy working on it.

A French mathematician called Henri Poincaré allocated set periods where he would not allow himself to be disturbed. In his case, 10 a.m. to 12 noon and 5 p.m. to 7 p.m. were the slots that he devoted to working solidly on his mathematical problems. Then, in the evening, he would read broadly around his subjects, perusing books and journals on a wide range of subjects as well as mathematics; "freestyle research," if you like. You would be surprised where the key ideas come from – it often has nothing whatsoever to do with the subject in question. Novels, art, lyrics – clues to the solutions to your problems lie in the most unlikely places. The key thing is to keep your eyes and ears open and keep looking for new connections. Like Archimedes climbing into an overly full bath causing the water to overflow – sometimes noticing the relevance of a spontaneous occurrence to a seemingly unrelated problem at hand can be just the leg up required to hit that *eureka* moment. This is the essence of creativity. Poincaré's greatest creative moment, when he solved a mathematical problem previously considered unsolvable, came to him as he was stepping off a bus. The A-ha! moment invariably occurs when you are thinking about nothing in particular. In the gap between adjacent thoughts the bubbles of original thinking are quietly breaking through to the surface.

## The night shift

When you sleep there's far more going on than just maintaining individual brain cells. You wouldn't believe how much work is involved when it comes to making, breaking and bolstering connections between cells. This is the basis of learning new skills and forging long-term memories. Test signals must be sent down the new circuits to ensure the correct cells are wired together after new protein mechanisms have been put in place to strengthen or weaken the influence of one brain wire over the next in line. You might be snoring away, but for your brain's infrastructure improvement workforce, it's all go.

This is where dreaming comes from: reactivating the same sensory brain pathways that have been used over the course of the day but, in the absence of light stimulating the eyes or sounds stimulating your ears, it is no longer properly anchored to reality – which is why such bizarre things often happen in dreams. When your brain is in sleep mode, it's far from idle. It's shifting up and down through the gears like a manic racing driver, it's just that it's in the simulator rather than out on the track.

## Rest and play

The problem for brains is that modern life doesn't allow people who work for a living much time for rest and play. The work bit, whether it's actual work or commuting to and from work or running around catching up with your personal affairs and keeping on top of day-to-day chores such as shopping, paying bills etc, really does get in the way of resting and playing.

The key thing to remember is that brains benefit from rest and play in equal measure, both are very important to them. And, when you get the balance right, it makes them far more efficient and effective when it comes to the work bit. Play enables your brain to lock into a less rigid mode of function than is usually allowed during periods of work. Instead of trundling through the same old set pieces, it is better able to invent new ones. Really useful, positive play for any brain involves its owner moving out of their usual environment, trying a new experience or activity somewhere else and, above all, really throwing themselves into the task wholeheartedly whilst being mindful of the importance of banishing all self-consciousness.

 The Corpus Callosum is a collection of 250,000,000 brain wires connecting the left and right sides of your brain.

## Release the prisoner!

*I spent six years researching phenomenally successful people and I discovered that they have all the natural ingredients of success that we as young children all once had in abundance – before we went to school! We, as curious preschool youngsters with vivid imaginations and no mental obstacles in our way, never stopped asking questions; but also we never stopped playing – no matter where we were or who was watching! Then, having entered through the school gates and become prisoners of other people's thinking, we had to grow up all of a sudden. The great news is that, deep down inside, we all still possess these childlike qualities.*

**– Adrian**

# Go jump in a few puddles!

By releasing the positive qualities we had as young children, including a love of playing imaginative games, our brains will experience much needed stimulation, driven by a need to be inventive. So, if you really do want to be radically innovative, you'll probably need to strip away a few layers of conditioning that were intended to make you "all sensible." Conditioning which, unbeknown to you, has entangled you in its net and restricted your natural urge to play.

It could be said then, that the success of an individual is largely down to having a ferocious appetite for answers, coupled with a childlike, unimpaired outlook. Having a hard working, conscientious, determined attitude is of course also vital. However, looking at it through the lens of science, we can see how important play is for problem solving and creative thought. Play entices your brain to process novel stimuli, activating areas that facilitate the process of finding common ground between seemingly unrelated ideas.

 Not everybody dreams in colour. Surveys have esti- mated that around 20% of people dream in black and white.

*Forcing your brain to process novel stimuli is the bedrock of creativity.*

## Chapter takeaways

- Are you receiving, over? Your brain is always coming up with solutions to problems and providing a constant stream of ideas. But are you tuned into them? If you're expecting solu- tions, make sure you're in a position to receive them.

- If you're going to give dunking a go, don't stay dunked too long – set an alarm.

- Your chances of resolving a problem will be greatly enhanced if you have got your subconscious on the case by inputting all the relevant data it needs to get to work on.

- Your brain loves and thrives on novelty. If you're in need of creativity, step away, get out of your box, do something different – preferably something that's random, novel and fun.

- When your brain is sinking into sleep mode, that's when it's at its most creative.

- The phrase "All work and no play makes Jack a dull boy" is still as true today as it was when it was first ever uttered.

# Perception Is Everything

## Now you see it, now you don't

The retina, at the back of your eye, is an extension of your brain. We experience the sensory world as if the sights our brain creates from the light striking the retina is a faithful representation of what is going on out there in the real world. In actual fact this is mostly an illusion. A very convincing illusion admittedly, but an illusion all the same. All sorts of short cuts and ingenious neural strategies are used to fill in the gaps wherever the brain does not have enough information to do a decent job of capturing reality as it really is.

Of course, under normal circumstances, the brain is so good at "faking it" that we are all under the distinct impression that what we see is accurate. But it's not always the case. What we see is actively created by a large, dedicated part of the brain, your Occipital Lobe situated right at the very back of your skull (see left side of the brain tube map for the Occipital stretch of the Visual Lines).

## Your permanent blind spot

Only the central part of the visual field is crystal clear. Every-thing else, believe it or not, is completely blurry. And, only the central part of the retina (Fovea) can see in colour, eve-rything else is in black, white and shades of grey. If you hold both your hands out in front of you at arms length, with both your thumbnails in the centre of your field of view, that is roughly the area covered by your Fovea. Only the Fovea has light detectors that are colour-coded and only in your Fovea are they packed densely enough to give you high-res vision.

So why does it feel like you can see detail and colour in the periphery? It's because your eyes are constantly darting around to capture high-resolution, colour information with the central Fovea. This lingers in perception for a short while to enable your brain to piece together an impression of the overall scene. These eye movements are so tiny and happen so fast that you are completely unaware of them. We are so adept at shifting our gaze to allow our Fovea to harvest light from an interesting blur in the periphery and then back again that we don't even realize it is happening. Your brain fills in the gaps between snapshots taken by the Fovea with an image that fits in with the overall scene.

If you are unconvinced then try this. Cover your left eye and with your right eye look at the O below. Then, slowly, move the page towards you. At some point before your nose touches the page the X will disappear. That blind spot has been there your whole life. Usually your brain effortlessly fills in this gap in perception.

O                              X

# Wiring up the senses

> During the first six months to one year, a newborn's brain creates 990 trillion new connections between neurons – consequently its Prefrontal Cortex uses twice as much energy as that of an adult's.

A newborn baby is not born with the ability to see as we adults do. The vision of a newborn baby is 1/20 of the resolution of adult vision – that's extremely blurry. The part of the brain that does what we think of as seeing actually "learns" how to see as the interlocking networks of brain wires are honed according to experience through the early weeks of life outside the womb.

Hearing is a different matter. An unborn foetus starts responding to sounds from the outside world from the third trimester onwards. So the connections between brain wires in the Auditory Line in the Temporal Lobe that create what we hear from the sounds that reach our ears have had a three month head start over those in the Occipital Lobe. We know this because one observant mother noticed that her unborn child would always change its behaviour the minute the theme tune to a famous soap opera started playing on the television. It's a similar deal with taste. Mums who snack on aniseed flavoured food and drinks when pregnant find that once their baby is born it has a preference for this flavour.

A surprising amount of progress is made in the womb when it comes to wiring up our senses and this process only accelerates in the first few years of life. After a child's brain has made headway in developing its abilities to sense and start moving around to explore their environment, somewhere between the age of 12–20 months, it starts to develop a sense of self. Strange to think that we had no concept of "I" or "me" throughout our first year. No wonder we don't remember any of it.

## Division of labour

Different areas along your brain's Visual Lines each have a complete map of visual space and each one extracts different aspects of any given scene. The brain stop known as V4 consists of a densely woven fabric of brain wires that actively extracts colour information from the light that hits the Fovea. We know that in rare cases where V4 is damaged in both halves of the brain, a person will lose the ability to see in colour. They see more or less the same patterns of both light and dark areas of objects, faces and landscapes, but it's all painted in shades of sludgy grey. This is because the job that area V4 evolved to perform is to compare the wavelengths of light in adjacent patches of retina and then "paint" the colour information into the visual scene created by all the other parts of the visual brain. Without V4 doing its job we simply cannot do colour. The fact is, outside your brain, there is no such thing as colour. Light in the outside physical world has no colour. What you see as colour is an interpretation made by your brain as it attributes colour to each wavelength of visible light. It is an illusion that helped us stay alive by making us better able to spot food and predators, but it is all in your mind!

At night our peripheral vision is better than vision at the centre. This is because the retina's high sensitivity light detectors (rod cells) are more concentrated around the edges than the middle of the Fovea.

Area V5 is located at the junction between the Occipital and Temporal Lobes and its job is to create the perception of motion when an object moves across your field of view. So, if area V5 is compromised in both sides of your brain, you won't be able to see moving objects. They remain invisible until they come to a standstill, at which point they will suddenly reappear as the intact areas of the visual brain kick in again.

Whenever you hear a particular song played on the radio the Auditory Line that runs through the upper part of the Temporal Lobe creates what you hear. There is strong evidence that a division of labour occurs in the sense of hearing as well. Some areas respond best to single tones (A1 stop), others prefer sound sweeps progressing from a low to high pitch or *vice versa* (PT stop).

A further division of labour was made evident when an advertising executive who had damaged a certain part of his brain selectively lost the ability to perceive music, yet his hearing of other sounds remained fine.

When you get a waft of a certain smell as you walk into a room the scent you perceive is generated in the inward facing part of the Temporal Lobe (Olfactory stop on your brain's tube map). Such sensory experiences are often associated with a certain person, place or time in life and can trigger an emotion and/or a specific memory which all come flooding back in an instant. Scents are particularly evocative in this way because the Olfactory Bulbs plug directly into the Limbic System unlike all the other senses, which go via the Thalamus.

## Bottom-up and top-down pathways

Over 50 years of neuroscience research has taught us about how "bottom-up" pathways take information from the eyes, ears, mouth, nose and skin to "higher" brain areas that actively create our perceptions. They first convert physical disturbances caused by light, sound, pressure, heat, liquid or gaseous chemicals into electrical impulses that your brain can analyze to create your impressions of the world. Visual pathways deliver these electrical pulses to the back of your brain where the information is fed into different patches of cortex, each of which crunch the data in different ways to produce different aspects of your visual experience. Similarly, the inner ear converts pressure variations that hit the eardrum into electrical messages that the Auditory Cortex (A1, PT, STG and STS stops on the brain tube map) can divvy up into different patches. These patches are highly specialized to analyze and actively create different aspects of the sounds we ultimately end up hearing.

There are also "top-down" mechanisms involved in sensing the world around us. These bring certain assumptions about what types of sights and sounds are likely to occur in different environments into the mix. As we gain experience of different places we can start to anticipate the types of sensory experiences that are typical in any given place and this dramatically speeds up the process of crunching the data. For instance, imagine taking a stroll in a European park at dusk and you hear a creature scrabbling around in the bushes but when you direct your eyes towards the noise you cannot tell what animal it is due to leaves blocking your view. In these circumstances top-down brain mechanisms will narrow the possibilities. You'll identify the mystery creature fairly quickly as a dog, cat, fox, bird or rodent according to its characteristic shape and movements, having automatically excluded the possibilities of tiger, rhino, kangaroo or monkey. Your sensory experiences are shaped by what you've sensed in the past, which has the effect of speeding up perception by filling in the gaps when the available information is incomplete.

## Sound the alarm!

Over the centuries, countless numbers of bleary-eyed sailors with the responsibility of keeping watch must have sounded the alarm only to discover that what they thought they had seen through their telescope wasn't in fact an enemy ship at all. With such a heavy burden of duty it's understandable why they would risk the wrath of the captain along with the entire ship's crew by getting it wrong, rather than risk the loss of everyone's lives; including their own. Naval technology may have advanced considerably but when it comes to making sense of weak sensory signals our brains are stuck with the same old kit.

*(Continued)*

During World War II, researchers found that radar operators performed completely differently in the context of training exercises as opposed to real-life combat situations. During actual combat there was a vastly increased incidence of false alarms, with operators concentrating so hard on the screens in front of them that they would often imagine blips on the screen that simply weren't there. These false alarms may have been a bit of an inconvenience but at least they rarely missed the approach of a genuine "incoming."

Conversely, in training there was no problem with false alarms but a far greater incidence of real blips on the screen that weren't spotted!

Expectation has an enormous influence on how sensory information is evaluated. During training the operator knows that a blip isn't really the approach of an incoming enemy. The consequences of missing a blip in these practice scenarios are minimal with perhaps a reprimand from an instructor or, at worst, a low mark on that particular exercise. The "response threshold" to use the official term, is higher than in a real-life situation – the decision to sound the alarm will only be taken if they are absolutely sure they saw a blip. On the other side of the coin, when carrying out exactly the same task for real, the radar operator knows that any blip could well be a potential enemy strike. The consequences of missing one single blip are unthinkable and so the response criterion is lowered. The outcome being that even the dimmest looking, fleeting, half imagined blip on the screen was likely to send everyone scrambling to battle stations.

# Do I know you?

Because of its immense importance, there is a specific piece of brain real estate that specializes in faces. The Fusiform Face Area's (see brain tube map) primary function is to process faces. As intensely sociable primates, faces are incredibly important to us humans. It is perhaps unsurprising to think that evolution has developed a dedicated patch for identifying and distinguishing different faces within just a split second, not to mention the expressions on them. Such information has proved invaluable to a creature like us. Over the centuries we have needed to be very quick at establishing whether an unexpectedly encountered face is friend or foe; it was often a matter of life or death. This ability was key to the survival of our ancestors.

However, when we come to associate certain faces with certain places, even this dedicated machinery can let us down. If a colleague we once worked with turns up on holiday then we can find ourselves completely unable to figure out why they look so familiar. The top-down mechanism that constrains which faces we usually encounter in which places, despite speeding recognition up under normal circumstances, actually slows us down when a face pops up out of context. We know they seem familiar but have no contextual cues to help resolve the conundrum of how exactly we know them. As we all know, this can be highly embarrassing, especially when they recognize you, start chatting and horror of horrors, start using your first name. You stand there, desperately trying to place them, not knowing whether to bluff it or come clean and ask them to remind you of their name. The crunch point usually comes when you suddenly find yourself in the truly awkward situation of having to introduce them to someone else and you're left with no choice but to confess!

Having a special area dedicated to processing faces can also lead to situations where we see a face when in reality no face is truly present. Seeing faces in the clouds for instance, or in the corner of the bedroom when we were six years old and scared of the dark. Low levels of illumination provide less information to the eye for the brain to use to define where one object ends and a separate object begins. In these circumstances such vague and incomplete information can result in the eerie sense of a ghost-like presence in a darkened bedroom. When our imagination gets the better of us (top-down), we can assemble the vague blurs formed by a variety of separate objects (bottom-up), and create the distinct impression of a monster's head staring at us (when in fact there is nothing really there but a dressing gown or perhaps an ornament on a bookshelf). Our brains are constantly looking for patterns in perceptual information and often find them even when they aren't really there!

## Mind the gap

Certain strong perceptions can end up being paired with potent associations. These associations can lead to what we usually think of as assumptions. Even a single word can lead to a whole set of assumptions. A stand out example being that of a major online retailer that had a button on their website with the word "Register" on it. On replacing it with the word "Continue" they saw an increase in revenue that year to the tune of $300,000,000. The word "Register" for most of us conjures up the assumption that we'll have to waste time doing things like filling in forms and having to type in information, whereas the word "Continue" gives us a feeling of progress, of quickly moving on, of getting somewhere.

## A deceptive encounter

*As a result of assumptions that get associated with our perceptions we all develop preconceived ideas about people; ideas that often prove to be wrong. Many years ago, whilst driving down from London to Somerset, I stopped off at a busy roadside cafe. Shortly after I'd placed my order a man that I can only describe as the biggest, scariest looking human being that I'd ever seen came in and took a seat at a large round table next to me. When he turned around to talk to the waitress, the words HELLS ANGELS were clearly displayed on the back of his jacket. He placed his order but, before it arrived, he quickly got up and disappeared into the toilets. Whilst he was in there, a small group of five, distinguished, older looking men wearing bowls club blazers came in. The only place for them to sit together was at the large round table, upon which its absent occupant's breakfast had now arrived. Seeing this, the five bowlers asked the waitress if anyone was sitting there and she replied "Yes, there is a gentleman sat there but I'm sure he won't mind if you join him." Obviously happy with this, they all took a seat. It may sound cruel, but I could not wait to see the reaction on the faces of these five old boys when what the waitress had described as*
*(Continued)*

*a "gentleman" came back to his table. When he did eventually return, I was not disappointed! On seeing the colossal "being" that was now suddenly towering over them, they all, without communicating a single word, jumped up, kicked their chairs back and shot off to find another table – as far away as possible.*

*Having had his breakfast my Hells Angel neighbour suddenly pulled out a copy of The Times newspaper and started finishing off what looked like an almost completed crossword. Noticing that I was watching he smiled at me and in a very soft, extremely well-spoken voice he said "Crosswords are always a bit of a struggle on a Monday morning!" We got chatting and it turned out that having left Oxford University he'd gone travelling, eventually ending up in Rwanda where amongst other things he'd been working with landmine victims. He was one of the most intriguing people I have ever met, an unforgettable, fascinating character. He was the inspiration behind my Pineapple Person character – from a previous book – those who are spiky on the outside but very different on the inside. How wrong you can be!*

**– Adrian**

Before we move on, a quick question for you. When you think of a wind farm, what picture springs to mind? Is it one of something that's clean, natural, positive and safe or is it one of something that's manmade, monstrous, noisy and damaging? Our associations with a concept such as wind farms can lead to a variety of assumptions that polarize our opinions – once formed they can leave us blind to the truth.

So, what has this all got to do with sorting your brain out? It's all about that gap between the bottom-up and the top-down mechanisms. This is where mistakes happen. But it is also where creativity is born. And the better you understand the quirks and shortcomings of this gap the less you'll misinterpret situations and the better you'll understand when those around you fail to "Mind the Gap."

We now know that information from the outside world is actively extracted by a wide variety of different brain areas (please see brain tube map – page 8) to create what you see (Visual lines), hear (Auditory line), smell (Olfactory stop), touch (Somatosensory line) and taste (Gustatory Cortex). These separate but highly interconnected places all do a very good job of summarizing what's out there in the world and help us get along in it.

Consider the mind-boggling complexity of the challenges facing your Occipital Lobe when having to conjure up a convincing 3D visual world from the 2D array of light-detectors at the back of your eyeballs. You can only begin to imagine the technical difficulties involved in capturing the sound of an 80-piece orchestra from a bunch of pressure variations in the air, or the finely-tuned senses needed to detect delicate individual flavours in a meal, the ingredients of which could come from an almost infinite selection of food sources.

But it is not 100% reliable. Your senses can trick you. People who are overly confident about what their senses tell them can often end up, having misread situations, looking and probably feeling rather silly. On top of this, the brain constantly makes assumptions about what is and is not likely to exist in any given environment.

## Impact of context and expectations on perception

Pleasant smells, foods and even music induce responses in the OFC brain stop – behind the middle of your forehead – which mirrors in real time the quality of your sensory experiences. We know this because neuroscientists scanned the brains of hungry subjects whilst they were fed bananas. Large responses in the OFC occurred during consumption of the first banana, reflecting the pleasure derived from satisfying their hunger. However, once they were stuffed silly and asked to eat yet more bananas, the additional bananas produced significantly lower responses, reflecting a decrease in enjoyment.

Other studies have indicated that the OFC does not respond to enjoyable perceptual experiences *per se* but rather how enjoyable you *think* they are going to be. OFC activations in response to a certain pungent odour were significantly increased when individuals were told that what they could smell was a high quality cheese. That is in comparison to when the exact same odour was squirted up their nose later on in the experiment, but this time participants were told it was a pair of sweaty socks. Your expectation can fundamentally influence your experience of certain sights, sounds, smells and tastes, positively or negatively via changes in the response of your OFC.

# TV experiment

*We took two bottles of wine, one an expensive bottle of grand cru, gold medal winning, limited addition Bordeaux and the other a very cheap looking bottle of mass-produced plonk. We poured the contents of both bottles down the nearest drain and, after rinsing them, refilled them both with exactly the same, average quality wine. Then we asked unsuspecting members of the public to taste the wines whilst wired up to portable EEG scanning electrodes to measure their brain's objective responses to these two "different" wines. Participants were also asked to describe the tastes and compare the two in terms of flavours, aromas and perceived quality.*

*Now what do you think happened? Did they cotton on to our little ruse? Did their sensory experience of exactly the same liquid tell them that they were being conned? Not a chance!*

*Their expectations, having been set up by the appearance of the bottle from which each glass was poured, along with an elaborate but completely fictional tale about each of these "two" individual wines unanimously tipped the balance in favour of the perceived "posh" wine.*

*Would your own taste buds be fooled by this simplistic charade? Or would you, completely uninfluenced by mere suggestion, quietly suggest that there has been some kind of mistake because both reds are in fact identical in every way?*

*It will come as no surprise that what they said about the flavours and aromas of each wine were totally different. The wine from*
*(Continued)*

the "posh" bottle was described in glowing terms and was deemed well worth the £40 price tag, whereas the wine from the "plonk" bottle was derided as being unimpressive, bitter and tasteless.

Our expectation of the amount of pleasure we might experience from a product can have a profound impact on how we actually experience it – so long as there are no powerful sensory signals that go against this expectation. If we had used the cheapest imaginable wine, alarm bells would have sounded and the experiment would not have worked. Equally had the wine from both bottles been a remarkable, exquisite wine with strong and distinctive character the experiment may also have failed as people would have found it difficult to believe that something so delectable could cost only £4. However, as there was no major disparity between the expectation and the sensory experience when the average wine was sipped twice – in the context of being poured from an expensive looking bottle and in the context of being poured from a cheap-looking bottle – their testimony regarding their perceptual experience of each followed their expectation precisely. Even more surprisingly, the objective measurements of activity in their OFC mirrored their subjective experience very closely. The more they said they liked it, the greater the response in the OFC as they sipped it!

**– Jack**

You have 12 million Olfactory receptors in your nose to detect different types of scent. Your average dog has at least 1 billion. Bloodhounds have 4 billion.

## Below expectations

Another glowing example of context leading to massive assumptions was when one January morning, during rush hour, a man started playing a violin at a Metro subway Station in Washington DC. After 43 minutes only a tiny handful of people, out of the 1,097 that had passed by, paused to listen to him before moving on, the vast majority appearing to be deaf to the talent within their midst. The sum total of his efforts amounted to just over $32. This busker was Joshua Bell, one of the finest violinists in the world who three days earlier had sold out at Boston's Symphony Hall, the average ticket price being around $100.

## Chapter takeaways

- Sights, sounds, tastes, touch and smells are actively created by your brain often from incomplete information – they do not always reflect reality.

- Mind the gap between your bottom-up and your top-down mechanisms – much "filling in" between the two can lead to inaccurate perceptions.

- Your brain is constantly looking for patterns and often finds them in perceptual information even when they aren't really there.

- Context can fundamentally change the way we perceive the world – fear in particular can play tricks on the imagination.

- Assumptions can change the way we interpret our perceptions – scrutinize your perceptions more carefully and you'll
start to see the cracks.

- Expectations fundamentally change our perceptual experiences – bear this in mind when you give or receive feedback
on what a perceptual experience might be like.

# To Do or Not to Do

## Icebergs of hindsight

Decisions, decisions, decisions! Everywhere you look there's a decision to be made. Faced with an ever expanding list of choices modern day brains are forever having to make them, some large, some small, some easy and some not.

Most decisions are made without you even realizing; with most of the action taking place deep down in your subconscious. Your conscious thinking on any given matter has an estimated capacity of 40 bits whilst the subconscious processing capacity is estimated at 11 million bits of information. In other words, the tip of your decision-making iceberg – the conscious part – is tiny in comparison to the immense bulk of information that is effortlessly crunched beneath the surface of your awareness.

The first *you* get to know about what's going on deep in the decision-making circuits of your brain is usually when, after much sub-surface deliberation, a conclusion has already been reached.

As a rational person, it really does *feel* like you are carefully considering your options, but there is ample evidence to suggest that this feeling is misleading. Most of the explanations you might give for why you made a certain choice are usually retrospective, the decision having already been made before you've even consciously mulled over the pros and cons.

Most of the time when you think you are making a logical, well thought out decision, you are really just attaching a feasible sounding explanation to a decision that you have already made on a purely emotional basis. So, in reality, instead of thinking something through, more often than not, you are unwittingly looking back in hindsight at a decision that your brain has already put to bed.

## Got a hunch?

With icebergs of hindsight in mind, and taking into account the huge influence of your well-developed perceptions, what you might regard as being "gut feel" decisions are in fact the result of an awful lot of behind-the-scenes hard work by your 24/7 brain. Thanks to its uncanny ability to learn from past experience, it's hardly surprising that when you trust your instincts, you are often proved to be right!

### Heads or tails?

Next time you flip a coin to make a decision, ask yourself when it's in mid-air, on which side up are you hoping "deep down" it's going to land.

Circumstances in which your gut feelings guide you best usually occur in situations where you have amassed extensive experience. If, for example, you have accumulated hundreds of hours of experience on the paintball battleground, your instincts will have become honed to know exactly when and where it would be unwise to suddenly stand up or go poking your head through an inviting gap, no matter how strong the temptation.

The instinctive feelings we experience in the pit of our stomachs when we make certain important decisions are, in fact, visceral sensations directly related to subtle emotional memories of the outcomes of past similar choices.

Instinct may at times feel literally like a "gut feeling." This results from blood vessels of the small intestine contracting to make more blood available for a busy brain. Your digestive system becomes a low priority when the brain is excited by or worried about an important decision. At these critical moments blood is diverted to your brain to meet its increased demand for oxygen and glucose. In other words the "gut feeling" that you can sometimes feel relates in part to the physical sensations produced when blood is siphoned away from your stomach to fuel your brain.

## Butterflies

We all at times experience "butterflies in the stomach." This fluttering sensation is a good sign. Its physical evidence that your Sympathetic Nervous System has kicked in to increase levels of alertness to help you perform at your best. When you are about to do something scary or exciting like delivering a speech or meeting someone you find extremely attractive, you can count on your nervous system getting you all
*(Continued)*

buzzed up and ready to go. This can all too often be perceived as a negative, but far from it. Any experienced public
speaker will tell you that it is actually a positive sign – their
performance can only really sparkle when the butterflies
come out to play.

In these high pressure circumstances, as well as blood getting
diverted to your brain it also gets dispatched to your muscles,
ready for action! The result being that so much blood is
squeezed out of the vessels that wrap around your intestines
that the tickly feeling of fluttering butterflies kicks in.

# Reward Line

Human brain imaging has confirmed that a specific brain network
swings into action when we feel pleasure – the Reward Line. As
decision making involves evaluating the potential reward value of
of each choice in the light of past experience, your "pleasure pathways" or Reward Line is vital to any decision.

The VTA stop (Ventral Tegmental Area) is at the very heart of your
pleasure network. It resides in your midbrain, an ancient part of
your brain sandwiched between your spinal cord – sending electrical signals in bundles of brain wires to and from your body, and
your Thalamus. This is the main biological junction box through
which these neural brain wires connect with the Cortex (the folded
outer surface of the brain), where most sensory and cognitive
processing takes place.

Quenching thirst, eating when hungry and sexual interactions all
induce responses in your VTA, not only making you feel good in
the moment, but also encouraging you to repeat the same behav-

iours in the future in the hope of re-experiencing the same pleasurable sensations.

The pleasure derived from receiving gifts, admiring a beautiful picture, listening to music or sharing a joke all stems from activation of your VTA. This in turn stimulates other parts of your Reward Line, like the Nucleus Accumbens stop, which is critical to the decision-making process as it's fundamental to our ability to anticipate future pleasures.

## What do you fancy for dinner?

There are five basic steps that your brain must go through during any decision-making process. Let's imagine, without taking into account any special dietary requirements you may have, that a friend gives you a call to invite you round for dinner and offers a choice of three dishes.

**Step 1.** You picture in your mind's eye all the options – in this case the choice is between a pasta, chicken or seafood dish.

**Step 2.** Your brain attaches a value to each option, a neural mark out of ten, if you like. Of course you are not consciously aware of how the exact details of this scoring system are worked out by your brain. The numbers are arrived at almost entirely at a subconscious level based on activation of the Nucleus Accumbens or "Buy Button." Neuroscientists hate this label because it is an oversimplification – but it is easier to say, and remember!

Decision making relies heavily on previous experience of the available options and the neural score that each alternative achieves is weighted heavily according to how much pleasure was derived from those dishes in the past. This acts as a prediction of how rewarding the option might be in the future.

Based on collective experiences, and to a large extent your most recent experiences, your brain attempts to predict how much joy is likely to accompany each option and settles on scores of 8, 7 and 4, respectively.

Let's imagine that, under normal circumstances, the seafood would have scored a predicted value of 9 rather than the 4 it got today, as it is your favourite food. However, a recent unpleasant seafood encounter has left you reluctant to go for this option and so your estimation of how enjoyable it would be has been drastically reduced, at least for the time being. Were you to be offered the same dish in a year's time, with the memory of the seafood disaster long behind you, it may well be marked back up again in the future.

Moments before you state your preference, you recall that you already had pasta for lunch; you don't really fancy it twice in the same day, so today its predicted value gets downgraded from an activity level of 8 to a 3. This leaves you with the chicken option topping the table with a sturdy 7. Despite being a bit boring (it's your default choice) at the end of the day you know you like it; it's a safe bet. Hence, the final values attached to the three choices end up: Chicken 7, Pasta 3, Seafood 4.

**Step 3.** Your brain simply compares these reward predictions. Again this occurs mostly just beneath the surface of your awareness, with the highest scoring option being the one you "fancy"

for dinner. In this case your "gut feeling" would inform you that you'd like your friend to cook you up a tasty chicken meal.

**Steps 4 and 5** are both made post selection and are vital to ensure that the decision-making process improves to serve you better in the future.

**Step 4.** This involves evaluating the outcome in light of the original prediction – was the chicken the best option after all? Or in retrospect would one of the other options have been better?

We've all experienced food envy in a restaurant where other people's choices look much more appealing than our own.

It is at this point that **Step 5** takes place whereby, having evaluated the outcome, we update the decision-making process for future reference. An error message is sent to brain areas involved in **Steps 1–3** to ensure that a better decision is reached next time round. For example, this may involve updating the predictions for this option in light of the fact that your friend's homemade "southern fried" chicken dish is rarely as tasty as the take-away version that you have become so accustomed to. Or, perhaps, if the meal was a spectacular success, the positive associations with this option would cause the likely reward value to be updated so that chicken moves up the scoreboard.

As our experience accumulates, we are, by the time we reach adulthood, all experts when it comes to everyday decisions like selecting what we want for lunch. Thanks to vast amounts of experience in getting this decision right or wrong we don't need to invest too much brain power to achieve our goal of biting into something that we know is very likely to satisfy. Consequently, we are usually pretty good at predicting which option will yield the greatest gastronomic enjoyment and can thankfully make such decisions on autopilot.

Playing classical music in a wine shop increased revenues by encouraging customers to purchase more expensive bottles of wine.

However, in many circumstances, we don't get to see the outcome of our decision until long after we've made our choice. Take for example choosing a holiday. The time lag between choosing a destination and actually getting to experience it, can often be several months. In these circumstances it can be very difficult to remember what you were thinking at the time you made the decision, which makes updating your decision-making process for next time round very tricky indeed.

This situation is highly problematic to anyone who wants to stop making the same mistakes over and over again. Under these circumstances most people rarely learn from their mistakes. One of the keys to many high-achieving people being so successful is that at the time of making a decision, they make notes on the main influencing factors behind their thinking. Once they know the outcome, if it's not the one they expected, they revisit their notes to get to the bottom of why things haven't turned out as intended. Despite the discomfort of going back, scrutinizing, being honest and facing up to their errors, they recognize it as an absolute must when it comes to staying ahead of the game.

The reality is that most of history's super successful individuals have all made a lot more mistakes than others – only because in comparison, they have had far more attempts at trying out new things. But what really sets them apart is that having taken on board lessons learnt from these past mistakes, they keep their decision-making systems bang up to date.

> *"Failure is simply an opportunity to begin again,*
> *this time more intelligently."*
> **Henry Ford**

# Emotion-flavoured decisions

Even with note taking, explicit memories of many individual decisions will have long ago faded from our conscious recollection, and in many cases will have completely disappeared. Forgotten they may be, but the feelings associated with the outcomes are still with us, safely logged away deep in the memory banks of our brains. They consist of a summary of the emotions that those individual experiences gave us – with the most poignant and the most recent examples being at the very top of the pile.

## Emotionally driven

*As a professional speaker spending a lot of time travelling, the car I use for running around on business is important to me. I have for many years driven an Audi, my current model being an A4.*

*These are the criteria that I consciously ran through in my head to arrive at this particular model of car:*

1. *Being a family man, as well as myself, I have to be able to fit my wife and children in too.*
2. *I'm looking for good performance and reliability.*

*(Continued)*

3. *I need to be comfortable on long journeys so that I arrive at events feeling reasonably fresh.*
4. *It has to be a car that I feel safe in – my business depends on me being able to deliver.*
5. *I do a horrendous number of miles each year so it really helps if it's economical.*
6. *I didn't want an enormous capital outlay, with my annual mileage it will rapidly depreciate in value.*
7. *It needs to be fairly small to squeeze into tight spaces. Often when I arrive at events these are the only ones left.*
8. *Image: I inspire people to be successful – I don't want to be seen turning up in an old banger, but at the same time I do don't want them thinking that I'm making too much money!*

*But why an Audi? After all there are plenty of other good car manufacturers out there that could supply me with a car that fits all my criteria. For years I've wondered why it is I like Audis so much but I've never been able to give myself a logical answer. That is until just a few weeks ago when one of my older brothers happened to mention our Dad's favourite old car, an Auto Union DKW – a forerunner of today's Audis. As a boy I loved that car. It was like no other car in the town where we lived and, as a working class family, we used to feel like kings of the road in it. It was the best car ever! Fond memories came gushing back on being reminded of it. The emotions are as real now as they were all those years ago. As I write this I can vividly remember the smell of the seats along with the lingering blue smoke from its two-stroke engine! Now I know the true underlying motive that drives my preference for the Audi brand. My love of Audi may be based purely on an autobiographical, emotionally rich child-hood memory, but it's been happily justified for decades with my "reasonable," and "logical" list.*

**– Adrian**

If we really wanted to understand why our brain is urging us to make one choice over another we may, in many instances, have to work back until we arrive at a poignant experience that happened long ago. At the time it may not have seemed like an experience of any note but it gave birth to a deep-rooted belief. A belief that always generates the exact same emotional bias whenever we think of it.

## Push and pull

When it comes to emotional bias, our more difficult choices are often finely balanced. We are drawn by a very subtle emotional *pull* towards similar options that have in the past resulted in positive outcomes and a gentle emotional *push* away from those that resulted in a negative one.

We've already heard that the *pull* appears to be generated in the Buy Button (Nucleus Accumbens stop). The *push* can be caused by several different factors, including the discomfort associated with a hefty price tag. This one is created in an island of brain tissue known as the Insula – a stop on the Limbic Line buried deep between your Temporal and Frontal Lobes. Another form of *push* instinct sets warning bells ringing in your Amygdala (also on your Limbic Line) when you worry that a choice is simply too risky.

## Alarm activated

Alarm bells can be set off by the slightest of things. If a yogurt is described as "90% Fat free" – not a problem. But if the same statement is reversed to read "10% Fat" – DING DONG!

*(Continued)*

This phenomenon is known in the world of psychology as "framing." You can describe any set of options in two different ways and dramatically shift people's choices. It all revolves around whether the options are described with a focus on the positives or the negatives. Both "90% Fat free" and "10% Fat" are accurate but they stimulate very different brain pathways. Whilst the positive perspective stimulates your Reward Line, the negative perspective stimulates your Amygdala. Usually sales environments will always focus your attention on the positive, but if you do want avoid being suckered, it's always worth mulling it over from the opposite perspective.

Overall this system of push and pull has served us surprising well over many years. It's clumsy in certain circumstances but overall highly effective when it comes to staying alive! Even in this day and age when survival is relatively straightforward, it operates perfectly adequately in most decision-making scenarios.

Our instincts can be trusted even when making highly complex decisions. When there is an overwhelming amount of information to consider, following our instincts can lead to good decisions, but only if we first invest the time to gather together and unhurriedly consider all the relevant facts. The secret is to postpone the final decision for a short while, to sleep on it for a couple of nights – giving your subconscious a chance to stir the pot – and only making the decision based on gut feel once it's had a chance to assimilate and make sense of everything.

## Danger zone – excessive buy button activity likely

Anything that makes us feel excited drastically increases activity on the Reward Line, bringing us closer to "going for it" – whatever *it* happens to be.

Danger zones are entered when activity on the Reward Line is seriously ramped up by exciting sensations, for instance those generated in a lively bar. Loud music, flashing lights and attractive bar staff, all help to drive up sales by throwing our Reward Line into a frenzy – making us more likely to spend far more than we ever intended.

In such circumstances the desire for instant gratification will prove almost irresistible, and any long-term plans of saving for a new car or summer holiday will go straight out the window.

*"The chief cause of failure and unhappiness is trading what
you want most for what you want right now."*
**Zig Ziglar**

The same fate has befallen many a determined plan to lose weight.
When in a restaurant, having stimulated the Reward Line with good
food, a few drinks and pleasant company, diets go out the window.
We eat 25% more food when dining with one other person, 39%
more when dining as a threesome and a whopping 79% more when
eating with three others! All it takes is just one person to ask for the
dessert menu and its goodbye healthy eating regime.

## Losses loom larger than gains

Brain scanning experiments have revealed that our favourite
part of the decision-making apparatus – the Nucleus Accumbens
(NA) – not only increases when the gain a person anticipates is
exceeded, but also decreases when the expected gain doesn't
happen. Worse still, if they suffer an unexpected loss the activity
level decreases even more. The most important thing to take into
account about the NA is that an unexpected loss results in *much
larger reductions* in activity than the increases in activity that occur
when there are gains of exactly the same size. Your Reward Line
overreacts disproportionately to losses in comparison to gains

This feature helps to explain why we humans are, on the whole, so
extraordinarily loss averse. Most people won't accept a gamble
unless the potential gain is at least twice the size of a potential
loss. Hence the minimum return in a casino is double or nothing.

Happily the reverse is also true – your NA is hyper-responsive
to unexpected gains but this time in the positive direction – so
you can use this "Power of Surprise Rewards" to your advantage. If
you give someone a present in circumstances where the recipient

has a fair idea what the gift is likely to be and when they are likely to receive it – like receiving yet another bunch of wilting flowers from the local garage on Valentine's Day – only a small increase in activity will be triggered in the reward pathways. This corresponds to the recipient feeling "quite" happy, but hardly overwhelmed. Not great, but far better than a fully expectant person not receiving anything at all – which would result in a rapid deactivation of the reward pathways and a corresponding crushing sense of disappointment.

If, however, there is an element of surprise in the gift-giving scenario then that exact same gift can induce a completely different response in the recipient's mind. If flowers are received completely out of the blue then there will be a disproportionately large response in the reward pathways and a commensurate surge in happiness; usually resulting in a rather large haul of brownie points!

> Giving is its own reward. Usually if you lose money, reward pathway activity is reduced, but if that money goes to charity, activity in these brain areas actually increases – making you feel good.

The main thing to remember is that the Reward Line creates a model of what is likely to happen on any given day, in any given environment, and these expectations are always adjusted and updated according to experience. The more an experience deviates from expectation, the greater the response in the reward pathways and the greater the emotional impact of that experience.

## The price of impatience

If you were offered the choice of being given £100 today or £110 this time next week, which would you go for? Many studies have shown that the vast majority of us would choose to take the £100. Our love of instant rewards and our fear of uncertainty have proven to motivate people to accept less value in return for immediate reward. And why not?! We all know that a lot can change in a week, the offer may be withdrawn, the person making the offer might keel over, or it might be us – the expectant recipient – who is no longer around tomorrow. Who knows what's around the corner! The problem with this tendency is that many of life's most important decisions, like saving enough for retirement, require immediate gratification to be snubbed in favour of much more important long-term goals.

## Never type angry

*You've probably heard the advice "never type angry." This is because when we're in any highly emotional state we make bad decisions, particularly when it is a negative emotion like anger. I strongly recommend that you DO TYPE angry but DO NOT under any circumstances SEND IT. Save it as a draft. Come back to it the next day and ask yourself: "What exactly do I hope to achieve by sending this communication?" There, right in front of you, will be all the evidence you'll ever need to convince yourself of the terrible decisions we make when in a negative state of mind.*

*Having re-edited the message several times over several days, only when you can, with hand on heart, re-read it without feeling the slightest hint of negativity should you actually send it. This is amongst the best advice I have ever been given. Thanks Dad!*

**– Jack**

# Marshmallows

A classic experiment carried out with four-year-old children makes a very strong argument for the benefits of resisting temptation in the short term for a better long-term return. It even suggests successful strategies to help in this endeavour. Each child was sat in front of a table upon which was placed a single marshmallow on a plate. Their task was simple: wait a few minutes whilst the psychologist leaves the room and, upon their return, if the marshmallow is still on the plate they will get a second one. Some of the kids just couldn't help themselves and gobbled it up before the psychologist returned. Others demonstrated incredible resolve and despite clearly being tempted, they managed to hold off and claim the bigger reward. This simple experiment had some profound implications. These same children were tracked for over 20 years. Those who could use their Prefrontal Cortex to rein in the impulses generated in their highly excited Reward Line and resist immediate gratification went on to get better grades in their school exams, got better jobs and earned more money than those who could not.

What we can learn from these kids is the difference in strategies they used to deal with the perceptual information reaching their brain from the outside world. The kids that *couldn't* resist the urges stared temptation in the face, looking lovingly at the enticing sweet treat. Those that *could* resist the urge directed their attention elsewhere. They shut their eyes and sung themselves a song. They turned sideways on their chair to find something else to look at, to make the time go faster. They showed an instinctive awareness that they

*(Continued)*

needed to take measures to reduce the sensory stimulation of their reward pathways and their Prefrontal Cortex dreamt up a solution to directing their attention elsewhere. We can all learn to control our impulses better. We just need to use a little imagination to find ways to reduce the excitement levels in our reward pathways when we're confronted with the lure of instant rewards.

# Experience really does count

Our instincts are right in many circumstances, but they can also be very misleading. Subconscious brain processes furiously working away beneath our awareness will do all the hard work of sorting all the relevant information for us, but they will only do a good job if we have extensive experience of similar situations. If we've never made a certain type of decision before, our intuitions will likely lead us astray – we'll always have a hunch one way or another even in situations that we have little understanding of. Our instincts can only guide us well once our subconscious has something concrete to work with and the more experience, the better the job it will do.

Buying more time – to investigate further in an emotionally neutral state of mind – is vital to making better decisions, especially when you are in emotionally-charged danger zone situations. Sometimes we need to take time to cool down because high levels of both positive or negative emotion usually result in the wrong decision being made. The context in which decisions are made, the brain state you are in at the time of the decision, can fundamentally alter the choices we make. Sometimes we do need to step back, look at things from a different perspective and to see that although the associated risks are worrying, or the long-term goals boring, they are the options that sometimes offer the best returns in the long run. It doesn't always pay to let your heart trump your brain.

## Chapter takeaways

- Your brain operates a neural scoreboard of likely outcome predictions based on recent and peak past experiences.

- Logical reasoning often justifies our emotional choice retrospectively. Identify past experiences that might make you feel that way – are they relevant now?

- Trust your instincts in circumstances where you have extensive experience, ensuring you're not in an overexcited state of mind.

- Distrust your "gut feelings" in circumstances where you have little experience or when a certain choice gives you a small quick win over a larger return later.

- Develop your skills of resisting the urge for immediate gratification; it'll pay off in the long run.

- Buy time – but don't postpone forever. Think, research, mull, *then* go for it!

# Food for Thought

## Don't wait for the leptin

Your Enteric Nervous System is affectionately referred to as the second brain; it is primarily responsible for regulating important aspects of digestion. It uses electrical circuitry running along the entire length of your intestines to coordinate the timing of rhythmical, muscular contractions to keep everything moving along.

Electrical messages are also sent to and from your brain via a large bundle of neurons called the Vagus Nerve. Your gut even has its own set of hormones that it uses to communicate via chemicals travelling in the bloodstream up to your brain. These hormones switch the hunger centres of your brain on and off according to whether your digestive system is empty or full.

Ghrelin is a hormone released mainly from your stomach when it is empty. It travels up to your brain's Hypothalamus to produce feelings of hunger, making you want to eat. Having eaten, when food starts moving from your stomach into your small intestine, another hormone is released – cholecystokinin (CCK) along with leptin which is produced by fat cells. Leptin and CCK have the exact opposite influence to Ghrelin. They make you feel full up and switch off your desire to eat more.

The only problem with this system of chemical communication is that there is a small design fault that is the cause of an all-important time lag. It can take a full 15–20 minutes after consumed food first expands your stomach before the hormonal signals reach your brain and even longer before that feeling of being full up kicks in. This delay is the reason why so many people carry on eating long after their stomachs are full. The inevitable outcome being that they end up feeling completely stuffed. This bloated feeling can have a big effect on our moods. As well as feeling sluggish and lethargic, it can make us feel a little disappointed with ourselves because despite having overindulged on so many other previous occasions, we've gone and done it again!

## Eat like an Okinawan

There is one place in the world where a greater proportion of people live to be a hundred years old than anywhere else – the Japanese island of Okinawa. As you can imagine, many researchers have tried to unlock the secret as to why Okinawans enjoy such long life expectancy.

Their diet is healthy, consisting of small amounts of meat, fish and lots of low calorie, but nutrient-packed fruits, seaweeds and staple vegetables like sweet potato and tofu. Oily fish is packed with omega oils – extremely good for brain health as they play a vital role in keeping neurons flexible.

In one experiment, when people switched to an Okinawan diet the men lost 18% of their body weight and the women 10%. Blood pressure came down by an average of 20 mmHg, blood glucose/insulin by 30% and cholesterol counts dropped from 195 to 125.

Similar diets, however, are eaten elsewhere in places where they don't age so well. Healthy diet is extremely important, but it is not the only factor.

A big clue to Okinawan longevity could lie in their dinner time mantra: "Hara Hachi Bu" which roughly translates as: "Eat until you are 80% full."

It's a no-brainer. If you don't overburden your digestive system, fewer substances that pose a long-term health risk will accumulate in your body.

**Tip 1:** In one sitting never eat more food than you would be able to hold in your two hands cupped together, the exception being a leafy salad. In which case, best make it two lots of cupped hands!

**Tip 2:** Large plates and large bowls promote the dishing up of large portions. This, combined with a tradition of encouraging people to eat every last scrap of food on their plate, ensures that an awful lot of people do an awful lot of overindulging. If you regularly eat too much then here is a simple remedy – get rid of all your old crockery and replace them with smaller plates and bowls. It may sound somewhat extreme but there is plenty of evidence to suggest that smaller crockery drastically reduces overeating.

# The influence of bacteria

Who'd have thought that the bacteria in your stomach could powerfully affect your mood? These single-celled organisms residing in your belly outnumber the cells that make up your entire body ten to one. It's been known for some time that these residents of your gut are essential for proper digestion. However, evidence that these self-same bacteria can determine your frame of mind is a very recent revelation.

The question is: how in the world could microscopic creatures in your gut possibly influence your brain? Well, for one thing, your gut's 100-million-neuron Enteric Nervous System is forever busy marshalling 10 trillion gut microbes to try to maximize your physical and psychological wellbeing.

In rats, where most of the research into the impact that gut bacteria has on behaviour has been focused, the balance between health-promoting good bacteria versus the disease-causing bad bacteria can be tipped by even the slightest amount of stress. Changes in gut bacteria have been found to affect not only physical health, but also pain perception and emotion, along with the variation in stress responses.

In humans, investigation into this exciting new area of research is in its infancy. One study found that when healthy human volunteers were given a 30-day course of "good bacteria," consisting of a mix of two different probiotics (Lactobacillus helveticus and Bifidobacteria longum), a decrease in symptoms of depression and anxiety became apparent. Another more recent study used the brain imaging technique fMRI to investigate the impact on meddling with gut bacteria on brain function. They found that in people regularly fed a yoghurt drink packed with probiotics, as opposed to those fed with yoghurt not containing probiotics, significant changes in the brain areas involved in creating emotional states were observed. It's early days, but all the evidence so far points to gut bacteria having a significant influence on the brain and, as a consequence, your mood.

Despite this evidence, the question still remains: how exactly do these bacteria do it given the physical separation between gut and brain? Gut bacteria are responsible for manufacturing 95% of sero-tonin – a neurotransmitter as vital for stabilization of mood as it is for healthy gut function. Gut bacteria also produce and respond to several other chemicals involved in nervous communication includ-ing acetyl choline, dopamine, melatonin, GABA and noradrenaline. You'd be forgiven for thinking that gut bacteria are able to influence your brain simply because of the chemicals they produce finding their way into your bloodstream and travelling up to your brain.

However, what gets into the brain from the blood is tightly regu-lated by a protective wrapping around the brain's blood vessels called the blood brain barrier. It seems that the main route of com-munication from gut to brain in terms of mood manoeuvres is actually via the Vagus Nerve. This cluster of wire-like neurons connect the brain to many different organs, including the lungs, heart, liver and gut. It enables them to be switched into "action stations" or "rest and digest" mode, depending on what is most appropriate at any given time.

In experiments with rats, when the Vagus Nerve is cut, the influ-ence of gut bacteria on the brain disappears. Emotional behav-iours, pain perception, stress response, all return to normal. Vagus Nerve stimulation is actually a last resort option for treating chronic depression in humans. What's more, it often seems to work.

Irrespective of the precise mechanisms by which they achieve this, your gut bacteria are something you should give serious consideration to. If you're feeling irritable, stressed or on a bit of a downer, you may wish to tip the balance in favour of your gut's "good" bacteria by providing reinforcements – in the form of pro-biotic yoghurt drinks!

## Fat filled and all sugared up

Let's take a look at the fuel that we should be putting into our brains to help them work better; the last thing we want to be doing is putting diesel into a petrol engine.

Although your brain weighs only 2% of your overall body weight, it consumes up to 20% of all the oxygen and sugar available in your bloodstream, and that's when it's just ticking over.

When it's working flat out, such as when you're really concentrat-ing hard on something, its demand for energy resources from your blood shoots up to 50%. Part of the reason you feel the urge to snack whilst deeply focused is because your brain is using up so much energy. What you choose to put in your mouth to try and meet its demands plays a critical role in how well it functions, not to mention the impact on your waistline.

More often than not people have a tendency to snack on foods made almost completely from sugar and/or fat. These irresistible, beautifully packaged sources of nourishment may feel like they are providing you the consumer with a satisfying instant fix. The truth is, as far as brains are concerned, these quick fixes are nothing short of a nightmare.

The reason why people love these sugar-loaded, fatty foods so much is because we humans have evolved over many years to find such foods delicious – as they were once extremely important to our continued survival. Man's craving for these types of food goes back to times when food wasn't anywhere near as readily available as it is today.

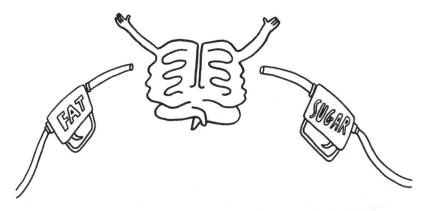

For the majority of people on this planet, death through starvation was once a very real possibility. When starvation is an imminent threat, it pays, for your own survival, to eat calorie-rich foods that are high in sugar and fat, which can be put into storage throughout your body for when food is scarce.

Life was once very much centred around simply staying alive and using anything we could lay our hands on to get by from day to day, a passion back then for consuming sweet, fatty foods made perfectly good sense. Our inbuilt sense of self-preservation hasn't dwindled one bit since then. Nor, unfortunately, has our deeply rooted, subconscious need to source foods that deep down we feel will help us to stick around a bit longer. It's only in recent years that we have begun to understand just how damaging these "comfort foods" actually are.

In today's developed world, the situation is very different. For most, the primary threat to health is not starvation but overindulgence in food and drinks that are always at hand and invariably high in fat and sugar.

Fast food outlets, kiosks and vending machines continue to pop up all over the place, making treats that look and smell more appealing and alluring than ever, easily available. Strategically placed, brightly lit and designed specifically with temptation in mind, peckish passers-by with a soft spot for immediate reward are drawn in like moths to light.

More often than not the companies that stock them are multinationals who know only too well how irresistible and addictive their tantalizing goodies are, especially to those on the move, in a hurry or just bored. These days it's an around-the-clock snack fest out there with on-impulse fixes available almost anywhere and at any time.

Sadly, it's something that has become the norm for many of the younger sugar-holic generation, with many developing their bad snack habits at an early age. That break-time fizzy drink and a packet of crisps that later on in life becomes a well-deserved biscuit with our coffee for "elevenses," a mid-afternoon chocolate bar, an after-dinner dessert and a ready sweetened, fat-saturated bedtime drink to round off a tiring day.

## Sugar-coated vandalism

For most, snack habits develop not necessarily because they are weak-willed or particularly greedy but because it all seems so harmless. In the 1950s and 1960s smoking seemed harmless. Everyone had an old aunt or uncle that they could point to who despite smoking a hundred cigarettes a day, was fitter than ever. It seemed like everyone was at it and they weren't dropping dead, were they?

Well, yes they were – but it wasn't obvious to the general public what actually caused the premature deaths of the vast majority of smokers. Doctors however had noticed that smokers seemed to die early and in more cases than not, from the same set of diseases. By the 1970s and 1980s the evidence that smoking caused a variety of cancers became overwhelming.

Sweets, fry ups and cakes might not give you lung cancer but high sugar-high fat diets are killing people! Here's why . . .

When you eat or drink foods high in sugar, for example fizzy drinks, sweets and cakes, all that sugar gets dumped straight into your bloodstream, quickly ramping up to potentially damagingly high concentration levels. As any diabetic will tell you, high sugar levels are bad news.

From a brain's point of view too much sugar is a big "no-no." You see, in too big a quantity it really does mess things up. It's no exaggeration to say it vandalizes your brain's lines of communication with your body. If you want to see how with your own eyes, go and make a meringue.

The special wrapper that helps your nerve fibres to carry messages from your brain to your body and back at lightning speeds is made almost entirely from fats and protein. In meringue making, when you add sugar to egg white a chemical reaction occurs that turns all that soft, gloopy, transparent protein into hard, white, chalky meringue. Delicious when served with a dollop of lemon curd and lashings of whipped cream but not so great when the sugar has made meringue of your brain wires!

In order to avoid the catastrophic consequences of having all your wires turned into useless strands of meringue, your body releases insulin. Insulin is a fiendishly clever hormone that reduces the concentration of sugars in your bloodstream. It does this by transporting the sugars out of your blood and into the surrounding tissues where they get converted into glycogen for storage in your liver and muscles until needed in leaner times.

## No vacancies

All well and good so far, but when the "no vacancies" sign goes up (in other words, when there is nowhere else left to store glycogen), that's when things literally do begin to go pear-shaped.

Running out of options and desperate to find a home, glucose starts getting converted into fatty deposits and seeks refuge under your skin. Not as you can imagine the most attractive thing to happen but nowhere near as dangerous in comparison to what also starts happening. Fatty deposits start getting deposited around your vital organs. Now that, without overstating it, is bad news.

So, if you've started to notice a few blobby bits about your person, there are also likely to be more blobby bits around the important bits that you can't see. The good news for people trying to shed a few blobs at the gym is: if you don't see any changes on the outside to begin with, don't lose faith, the main threat to health – the fat around internal organs – is the first to be shed.

What it all boils down to is that if you are a serial snacker, all the excess sugar you snack on is converted into fat because all the spaces in your glycogen stores are more than likely to be fully occupied. On top of all this, eating fast-release sugar snacks stimulates an all-too-sudden release of shed-loads of insulin. This dramatic hormonal release causes tissues to suck too much sugary glucose out of your blood, which is why you feel really hungry again just a short while after you've eaten. It's a simple rule, if you're producing too much insulin as a direct result of eating large quantities of sugary carbohydrates – you'll very soon be running low on sugar.

# Sugar rush – no myth

A sugar rush does make you feel good but it's not so much to do with the impact of the excessive sugar but more to do with the emergency response of the large dose of insulin released by your pancreas into your blood. Insulin is released to remove as much sugar as possible and to help prevent you having a whole range of health problems.

The problem is that insulin doesn't just extract sugar from your blood, it also takes out amino acids – the building blocks of proteins. That is all the amino acids apart from one that is extremely important to your brain – tryptophan.

Tryptophan is the basic essential building block of a variety of neurotransmitters including dopamine, a vital component of the reward system that makes you feel happy. Under normal circumstances tryptophan has to compete with all the other 19 different types of amino acids, queuing up at the tightly controlled entrance to your brain – the blood brain barrier. But with the insulin sucking the competing amino acids out of the race the tryptophan can easily stream into your brain with fast pass access. The high availability of tryptophan leads to greater amounts of dopamine and sero-tonin in the brain, which is why you end up feeling good.

Sugar rushes may be an effective way of making yourself feel good in the short term, but in the long term they have the potential to destroy your health. They will make you fatter in the medium term as sugar continues to be converted into fat, and after a lifetime of asking too much of your insulin-blood-glucose regulation system, the long-term prognosis is diabetes along with a whole range of spin-off health problems.

# The best fuel

Avoid fast-release sugars and instead eat slow-release sugars. This is the only way to avoid the "sugar roller coaster" – rocketing and plummeting blood sugar levels which are commensurate with periods of high energy and manic behaviour being quickly followed by low energy, low motivation and irritability.

Slow-release carbohydrates like oats, vegetables, whole grains, gradually release sugar into your bloodstream over a period of several hours, so there is no need for a massive release of insulin. For example, having porridge oats for breakfast for the slow-release sugars to help keep you going until lunchtime, perhaps topped with apple, banana or berries to give you a bump start, instant energy boost.

When the urge to nibble does become irresistible, drink water, eat fruit, vegetables or wholegrain snacks. Being in a hungry state with all that Ghrelin and other appetite-stimulating chemicals swilling around in your brain fundamentally changes the way you make decisions. Deciding what to eat when very hungry compared to a bit peckish has a big impact on how much food you try to eat. Not being so hungry at lunchtime and less likely to overeat will mean you won't end up feeling like you're carrying around a "food baby" all afternoon. Food babies like all babies need looking after. Not only do they make you feel completely stuffed they also make your brain all sluggish as blood is diverted from it to your gut to take care of your demanding baby.

If you want to stay on the ball, it's the only way to go. Slow-release snacking means you are full after a modest rather than a large lunch, which means you can remain energized, more alert and hopefully more effective. Instead of going from one moment positively buzzing and the next dropping down to feeling lethargic, unable to focus and getting all irritable, snacking on slow-release carbohydrates ensures that your energy levels are sustained on a more even keel. Throughout the day you'll probably feel a lot happier within yourself, your mood swings won't be so dramatic and those around you will also probably benefit from your new-found slow-release carbohydrate diet!

## Three quick points

1. Sugary snacks do have their place – you just need to use them sparingly and strategically. Prior to a big perform-ance it doesn't hurt to eat something packed with fast-release sugars. If your adrenaline is really pumping then your metabolism will be high and the sugar will get used to give you the energy you need to thrive.
2. Healthy snacks are healthy if eaten in moderation. Just because they are "healthy," it doesn't mean that you should non-stop gorge yourself on them! A lot of cereal bars have more calories than a chocolate bar so select carefully when picking them up in the supermarket. Fight the temptation to eat more than one in any given hour.
3. Go bananas! Bananas are the perfect on-the-go snack. Amongst other things, they fill you up, they're loaded with natural sugars, they contain tryptophan, they are high in fibre, they even help prevent cramp and they come ready wrapped!

## Pump that brain

Forget body beautiful, as far as your brain is concerned the body is just a vehicle to help it get around and maintain a steady supply of essential chemicals. When people think about exercise, the emphasis always seems to be about the benefits to their body, giving little thought to their brain.

Today's media very much focuses people's attention on exercises designed to reshape out-of-shape bodies. But as far as your brain is concerned, losing fat and toning muscles are just superfluous by-products in comparison to the profound benefits it gains from you taking regular exercise.

The good news is that unlike the frustratingly long periods of dedicated training it takes to yield the slightest of visible changes to your body, there are several, noticeable and almost immediate improvements to your brain.

The one key thing your brain gets out of you exercising is more blood. It gets more because your heart pumps harder and faster. This occurs in response to a chemical released from the glands that sit on top of your kidneys – adrenaline – and also because of a set of brain wires that branch off your spinal cord to innervate more or less every organ in your body – your Sympathetic Nervous System.

These automatic hormonal and nervous system responses to exercise increase the speed and depth with which the lungs inflate and deflate – upping the intake of oxygen and removal of carbon dioxide waste. The adrenaline also causes sugars to be released from storage in and around your body. So not only does exercise mean your brain gets more blood passing through it, it also means that blood is richer in oxygen and sugars which are required to create the energy that keeps it firing on all cylinders.

Improving the supply of glucose-packed, oxygenated blood that surges through it is so important because, like all brains, yours has no room for storage. Virtually every cubic millimetre of the space inside your skull is taken up by 86 billion wire-like electrical brain cells and a further 86 billion glial cells that provide the support network upon which your whole communications infrastructure is built and is critically dependent upon.

Every single one of those cells rubs up alongside a tiny blood vessel – it's the lifeline to those essential energy molecules that it needs to stay alive. With nowhere to store the glucose and oxygen that combine to release the energy to keep your brain cogs turning, each cell is 100% reliant on a constant flow of blood via microscopically thin tubes that weave an intricate web throughout the massive ball of brain gloop sloshing around in your skull.

## Consistent high performance

Exercise is a tool that can be used to change your mood in an instant. Several powerful hormones (of which adrenaline is but one) are automatically released into your bloodstream when you do even a short burst of exercise exerting a positive effect on your mood. A 20 minute jog around the block followed by a quick shower is all that is required to make you feel much more alert and able to concentrate fully. The time invested in a short bout of exercise will be more than compensated for when your brain is positively zinging with everything it needs to function to the best of its abilities.

 There are no pain receptors inside your brain, so brain surgeons can prod around without hurting the patient.

For your brain's sake, try not to think of body builders pumping iron on the beach or Olympians performing superhuman feats – daily, gentle exercise is all your brain needs for boosting its performance. The great thing is, once you've got into the habit of doing regular exercise for just a few weeks, your general mood will improve. The blood vessels to your brain will be clearer than those belonging to someone whose only exercise is running for the bus; and your heart will be stronger, enabling it to pump blood through your brain more efficiently. These factors combined lead to improved mood by increasing the delivery of oxygen, but more importantly increasing the removal of waste materials constantly generated by your brain as a result of the fact that it is always switched ON.

 Up to 1 litre of blood passes through your brain every minute.

Exercise is probably best regarded as being a multi-purpose tool that has both an immediate and longer-term impact on increasing the efficiency of all your brain functions. It can give you the opportunity to mull things over away from work and, at the very instant your heart starts pumping faster, it can help snap you out of a gloomy mood.

## Endorphins – natural opiates

Key to a more positive mood change taking place is also very much thanks to endorphins being released when your brain detects an increase in bodily movement. The original reason why we evolved to release these natural feel-good chemicals in response to exercise is all down to their ability to shut off pain signals and promote survival in dangerous situations.

If you should ever need to run away from danger, fight off an enemy, or take on any other potentially life-threatening challenge you wouldn't want a twisted ankle or a stubbed toe impeding your escape. Endorphins are the reason why people sometimes walk away from accidents only to drop dead of their injuries a few hours later. Brains use them as a natural form of opium to numb inconvenient pain messages.

The endorphin release that accompanies moderate to intense exercise is literally designed by nature to make you feel high so that you won't feel the pain that would otherwise make you stop. The bonus is that these endorphins are still swimming around in your body when you get back from doing exercise and continue with your work. Available to anyone, completely free, ridiculously powerful and perfectly legal, endorphins are guaranteed to change your mood in a positive manner.

 Regular exercise can be as effective as drug therapy for stroke and diabetes management.

You may think that brains being brains are fairly streetwise, but the truth is that in these circumstances they honestly can't tell the difference between you genuinely being scared witless and fighting for life or you being on a treadmill in the safety and comfort of a gym. So the next time you want to feel on a bit of high, a good workout should do the trick.

## Chapter takeaways

- Try to stop eating a meal before leptin reaches your brain and flicks the off switch.

- Your gut bacteria really do have an influence on your moods. Good bacteria need all the support they can get.

- Don't go making meringue out of your brain wires, too much sugar only leads to vandalism.

- Make sure you're giving your brain the right fuel. Fill up with slow-release carbs every morning and when "topping up" in between meals.

- Remember, as well as your body, your brain also benefits enormously from regular exercise.

- If you want to feel great – do endorphins!

# All Aboard the Stress Express!

S tress is a friend. Without stress, you'd be constantly flummoxed, you'd never get anything done and you'd certainly never manage to rise to the challenges presented by an ever-changing world. Stress, or more precisely the stress hormones, cortisol and DHEA, are the keys to your ability to respond effectively to changes both within your own body and in the outside world. Without the ability to release cortisol and DHEA you would lack the behavioural flexibility that enables you to do every worthwhile thing you have ever done.

Stress, if controlled and harnessed, can provide many benefits but only if you don't let it hang around too long. Cortisol works best for you in short doses but too much cortisol over long periods can be extremely detrimental to your health. If you want to stay friends with stress and benefit from it, it's best that you don't let it outstay its welcome.

Stress in recent years has got itself a bad name. This is largely due to the fact that it is widely misunderstood. When faced with stressful situations at work most people respond by not only working harder but also by working much longer hours. Tired and irritable they gradually become more and more stressed. They start taking stress home with them, they take it to bed with them and they even take it on holiday. Lack of quality time spent with their family means that they too get stressed. With tensions mounting and tempers flaring, guess what? More time is spent at work to avoid conflict at home!

What has happened is they've unsuspectingly boarded an ever-accelerating, non-stop train with no control over its speed or where it's heading. Unable to slow it down or hop off, they just keep going, riding along on the runaway *Stress Express* – just to overcome a problem at work.

They have no idea as to what in their body creates stress, nor a clue as to why their brain is making them feel so stressed out. As a passenger, with no hand on the controls, they'll inevitably end up coming off the rails.

## Stress heads

Stress hormones are indeed a double-edged sword. A moderate amount can truly help to improve performance by mobilizing glucose for immediate use and switching your attention systems to action stations. Large amounts on the other hand can dramatically impair performance, even in the short term.

The importance of learning to control your cortisol levels cannot be emphasized enough here. Chronic stress – high cortisol levels over long periods of time – changes your body in a very unwelcome fashion. It can make you fat, can cause muscle wasting, high blood pressure and it hobbles your immune system, leaving you vulnerable to opportunistic infections.

These, however, are all fairly minor in comparison to what it can do to your brain. Chronic stress interferes with the connectivity in Frontoparietal brain areas that constitute your Hold Line. This reduces your working memory capacity and, as a direct result, your problem-solving ability. Worse still, it increases the synapse-covered dendritic branches in your Amygdala and decreases them in your Hippocampus – shrinking your memory banks.

This all increases fear and anxiety levels to the point where high risk is perceived even where there is, in reality, very little and causes a bias in memory recall towards negative emotional events. Ultimately this can lead to such extreme risk-aversion that often the best decision isn't made because it involves a small amount of risk. "Nothing ventured nothing gained" goes right out of the window. Only absolute certainty can be tolerated.

## Cortisol the motivator

The rather unpleasant feeling of being stressed is actively created by your brain in response to the presence of stress hormones. The relay race of chemical messages that results in the release of these stress hormones occurs when the brain detects a problem that needs urgent attention.

Your Hypothalamus sits on the underside of your brain and has a structure called the Pituitary Gland hanging off the bottom of it. When a stressful situation arises your Hypothalamus sends a chemical message to your Pituitary Gland causing another hormone to set sail in your bloodstream all the way down to your Adrenal Glands which sit directly on top of each kidney. When the hormone docks with the outer layer of your Adrenal Glands a manufacturing process is started which quickly results in the release of cortisol, DHEA and the other stress hormones.

Cortisol has VIP access to every single cell in the body. Unlike most of the other chemical messengers used in the human body, cortisol is lipophilic (fat loving). This means it can mix freely with fatty substances and so pass unchallenged through cell membranes. Most chemical messengers are hydrophilic (water loving), when they hit cell membranes they are unable to pass through. The fatty bubble of membrane surrounding each cell is how your cells keep their contents safely inside and how they keep what is in the surrounding fluids out. Cortisol however slips freely through all these membranes like a ghost through walls.

Not only can cortisol enter into every single cell of your body and brain, it can also get inside the nucleus of each of your cells, the nucleus being where all your DNA resides.

The 23 pairs of coiled double spirals, one set from your Mum and one set from your Dad, is where the genetic code for every single protein required to build a human being resides. Special receptor molecules on the DNA bind to any cortisol that happens to be knocking around inside the nucleus which causes many genes to be either switched on or off – like a sound engineer pushing sliders on a mixing desk when recording a song.

The impact of hormones released into the bloodstream is so profound because they effectively switch every cell that makes up your entire body into a different mode. By orchestrating which genes are switched on and off, cortisol puts your body and brain in a mode that's best suited to deal with, and ideally remove, the source of the stress you're experiencing. This is the point that people often miss when they get stressed about being stressed. They don't realize that a little bit of stress actually helps them get things done better and gets to the bottom of the problem that is causing them stress in the first place.

Even the mildly unpleasant emotional state of being stressed is helpful – it motivates us to act; to get up and actually do something about the situation that is creating the problem. We humans are highly efficient creatures, masters at conserving energy unless spending it is really necessary. To put it another way, we're lazy!

Discomfort is an extremely effective way of motivating us into action. If we didn't get stressed, cortisol wouldn't be able to work its magic and nothing would ever get done. Cortisol induces a variety of physiological effects that together provide us with more energy. In so doing it plays a key role in enabling us to adapt to change. Cortisol enables us to mobilize resources in response to internal changes such as an increased demand for energy from your brain cells when a work deadline is due to be met. It also enables us to respond quickly to external changes like getting an unexpected call informing you that a loved one has been taken into hospital or that at very short notice you've won a weekend away.

Your brain is the most important sex organ – it's what turns you on! Your largest sex organ is your skin – all 1.8 m² of it – the size of a single bed sheet. And sex happens to be one of the best antidotes to stress.

# Cortisol in action

At times we all worry and end up getting stressed out about things that we shouldn't. They are either: things that turn out to be figments of our imagination, or that are real but beyond our control. The reason for unnecessarily stressing is because of our brains inability to distinguish between what is real and what is vividly imagined – no matter how ludicrous in the light of day those imagined events may seem!

Your brain makes little distinction between what is real and controllable, and what it believes is actually causing perceived problems, often based on flawed assumptions.

As soon as you start to give something a lot of serious thought, whether it be real, imagined, within your control or not, your cortisol will swing into action. That's why when you suddenly wake up in the middle of a horrific nightmare your heart is pounding and your breathing is short and rapid.

Give this a go. Close your eyes and imagine that having just cut a fresh lemon in half, you're now holding one half of it in your hand. Bring this juicy half up to your nose and take a really big sniff. Notice how strong its citrus smell is. Now squeeze as much juice as possible out of it into a tablespoon and hold the tablespoon near your mouth in readiness to drink the contents. Now very slowly, without spilling any, bring the spoon closer and closer to your mouth. Get ready to drink, open your mouth – and stop! Chances are, your brain has ordered extra saliva be produced in your mouth in an attempt to dilute the citric acid – a real physiological response based on an imagined sensory scenario.

## Holding back the pain

As well as regulating the amount of available energy to meet changing demands, cortisol also has an enormous impact on your body's defence systems. Upon detecting an invading lurgy, your ever alert immune system will immediately spring into action. The trouble is, in using all the weapons at its disposal to fight it off, it will more than likely cause some discomfort, possibly pain, and at the same time make you feel unwell as your energy is sapped in the struggle to repel the uninvited invader. For example, in combating a throat infection your immune system's attacks on it will create swelling, making it painful to swallow and leaving you feeling under the weather.

In times of crisis, when you really do need to get on with something, cortisol temporarily holds back your immune system from getting fully stuck in until whatever it is you need to get done has been done. It sends a powerful message to your disease defence system that says: "Now is not the time for me to feel ill, I can't afford to – now go away!" Fighting disease with our vast armies of immune cells is an extremely energy consuming activity. By postponing your immune system response all available energies can be channelled into dealing with the pressing matter at hand.

It is hardly surprising then that performers, despite suffering from a multitude of ailments, manage to get up on stage, forget about all their aches and pains, and deliver great performances. It's also not surprising why most self-employed people will tell you that the best cure for flu is to go self-employed!

## Who's driving your train?

Travelling on the *Stress Express* is not a problem providing, that is, YOU are the one doing the driving. If you want to regard stress as

## Pathetic is good

The helplessness that we exhibit when we fall ill is actually a behavioural set piece that is switched on by our immune system to prevent us from wasting energy moving around and fretting over the concerns of everyday life. Feeling and acting pathetic when ill helps us to conserve valuable energy so that it can be invested in the singular pursuit of conquering invading bugs and infections.

a friend and fully utilize the many positives that it has to offer, you must be the one that decides when and where you slow down, when and where you stop and when and where you get off. In other words you and not the outside world must be the one in control, running your life.

Rest stops along the line are extremely important. There are several different categories of rest that you should consider – each of them vital in terms not only of reducing cortisol levels, but also permitting important cellular maintenance work to take place to keep your body and brain working well on the timescales of days, years and decades.

Taking small bite-size periods of time out throughout the day helps you get through each week better. Ensuring that your week-ends, or whatever days off you get, allow time for some periods of total rest will help you get through each month more effectively. As well as these, what really does help is if you take a stress-free holiday each year. It will allow your overworked brain to do some deep cleaning and seriously needed repair work. An uneventful or even boring holiday is perfect in this regard!

## Reducing cortisol with laughter

Laughter reduces stress by, amongst many useful things, reducing levels of cortisol. It is an invaluable tool when used appropriately during stressful interactions at work and at home. Making a carefully timed comment that causes people to laugh doesn't just reduce your own cortisol levels it also brings down levels in those around you too.

Laughter involves many different regions across your whole brain – particularly your brain stem and more or less every stop on the Limbic Line.

Periods of rest where you can fully unwind are essential if you want to ensure that your brain rises to major challenges in the months ahead. They will help to ensure they are dealt with to the best of your ability and without causing any long-term damage to your health.

Your brain may be the most sophisticated piece of bio-ware in the known universe but expecting it to work at its absolute maximum all of the time is just wishful thinking bordering on bonkers.

Like all engines, it needs maintaining and that takes time. Expecting it to carry out running repair work whilst working flat out would be like expecting someone to re-tarmac a high street in rush hour without redirecting the traffic. Imagine it: a convoy of lorries, vans, cars, bikes and pedestrians all in a big hurry to get from A to B whilst a maintenance crew are scurrying around between the vehicles trying to do some much needed road works. The best you could expect would be to fill the odd pot hole here and there. Well that's just what it's like for your brain when you're expecting it to run at 100% a 100% of the time and, at the same time, stay in tip-top, fully maintained, working order! If you want to get the best out of it, running repairs on the job will only keep it performing well for a limited period of time. Before long, it'll begin to go past the point-of-no-return and you'll slowly begin to conk out. If you really do want it to deliver on demand and put in peak performances over a period of decades, it needs time to catch up, it needs to rest and repair properly.

Whenever you stop moving around torn muscles can be rebuilt, food in the stomach can be properly broken down so that nutrients and building blocks can be absorbed into the bloodstream and old bone cells can be gobbled up by special immune cells to make way for stronger healthier new ones. On top of all this, repairs can be carried out on your organs to make sure they are able to function fully and efficiently perform their specific roles. This kind of maintenance is very important for your body, but it's *extremely* important for your one and only brain. On a day-to-day basis it will of course do whatever it can in the natural lulls, repainting the road markings and relaying cracked paving slabs, but you would be well advised to actively create restful gaps between your rush hours of mental activity, wherever possible.

## GOM Time

*Once or twice a day I take a couple of minutes out just to pause, step back, relax and focus on what lies ahead. I call it GOM Time, GOM being the Tibetan word for mediation. All I do is sit somewhere quiet and take a few deep breaths in and out whilst picturing whatever it is that I next need to be getting on with. I use a very effective, quick and well-known technique called "7–11 Breathing" whereby you simply breathe in whilst counting to seven and breathe out counting to eleven.*

*The key is to breathe out longer than you breathe in. I find it really helps to un-clutter my busy mind and it certainly helps me to be more productive. I use it to "power up" so I'm ready for the next big challenge or to "power down." For example: when I'm driving home having spoken at a conference, I pull into a lay-by just down the road from where I live and allow myself some GOM Time. I do this to try to bring my spinning head down from travelling at a thousand miles an hour so that when I do step through my front door it's going at a much slower pace and I'm able to enjoy precious time with my family. And, to help make sure there's no stress tagging along behind, I leave my luggage in the car and come and get it later, when I've completely unwound.*

**– Adrian**

## "Trying" to get to sleep

"Trying" to "do" anything involves different areas of your brain working together to perform a task. You cannot try to go to sleep because trying suggests effort, which is the complete opposite of what happens when you fall asleep!

"Allowing" your brain to "fall" asleep is a better way of thinking about it and the mind-set that facilitates this process is all about surrender. Getting frustrated that you cannot get to sleep is the worst thing you can do, because frustration is an active not a passive process. Instead of slipping into neutral, you're revving up your brain.

As well as the obvious ones like never drinking coffee just before going to bed, if you do have difficulty sleeping and are badly in need of a good night's rest, then there are several things you can do to help your brain. The first thing is to begin to think of your bedroom as an environment that must be completely free of any unnatural stimulation, especially electrical. If you watch television in bed just before you "try" to fall asleep you are making your brain more active at exactly the time when you should be making it passive. If you use the internet in your bedroom just before you sleep, again you're getting it all revved up and stimulated when it should be winding down and drifting into sleep mode by being bored into submission! The only activities that should ever take place in your bedroom at night are sleep and sex.

That way, after a few weeks of training, your brain will begin to subconsciously associate that particular room with peaceful slumber. Eventually, the moment you think about going to bed, even before you've crossed the threshold into your bedroom, it will start going through the processes of shutting itself down.

## Sixty crucial minutes

One thing you can do to help is make sure you don't do anything that requires your brain or body to do any work in the crucial sixty minutes before bedtime. Make sure you've eaten well in advance so that you are not all bloated and trying to digest a big meal when you get into bed. Make sure any exercise (apart from sex) is taken as early in the evening as possible so that all the stimulating adrenaline that gets triggered by it has had a chance to be cleared out of your system. And, if you shower or have a bath, also do that a good hour or so before hitting the sack.

The key problem with exercising or having a hot shower or bath before you sleep is that your brain needs to cool in preparation for entering the sleep state. Blood vessels in the hands and feet dilate to help radiate heat away to bring core temperature down, which in turn cools your brain. If you create a lot of heat through exercise, absorb it from bathing or, produce it in your stomach in the process of digesting food, it makes it difficult for brains to get into sleep mode. If you're not a sound sleeper – it really does pay to make the last hour of your day as dull as possible!

This one may seem obvious but it is a real sleep killer for most couples. Do not, under any circumstances, initiate heavy conversations about serious topics just before bedtime; it's much better to discuss them elsewhere in the day. Better still find a set hour at the weekend during which, on a weekly basis, you take turns voicing your frustrations, worries and concerns. Upsetting, distressing, worrying or emotional subjects obviously need, at times, to be discussed and there is never really a good time to broach them. However, the very worst time to lock horns over things such as relationship difficulties, family arguments and money is just before bedtime as it gets your brain activity and cortisol levels up when sleep needs them to go down. Avoid pre-bedtime stress-inducing conversations like the plague. Through getting a good rest, your brain will work a lot better for you the next day.

*The last sixty minutes of the day must, if you want to get a good night's sleep, be as stress free as possible.*

## Nocturnal rewards

*A few years ago I found myself constantly waking up in the middle of the night and was finding it almost impossible to get back to sleep. My lack of sleep began to have a huge detrimental effect on me. As I continued over a period of months to struggle with my reoccurring nocturnal problem, I was finding it increasingly hard to concentrate at work; I lacked energy, felt lethargic and became increasingly short tempered and ratty with people around me. It became, and please excuse the pun here, an ongoing, living nightmare. I used to do a whole variety of things to try to make myself feel tired and hopefully get back to sleep. I'd sit up and read books, go and get a drink, watch TV, make myself something to eat, take a bath or even on some occasions, go out for a middle-of-the-night stroll. I'd do practically anything if I thought it was going to result in me being in the same enviable position that the rest of my family were in – fast asleep!*

*None of these things I tried ever worked. Then one day someone gave me an absolute gem of a tip. They explained that all the things I was doing to try to get back to sleep were the very reason why I was waking up in the first place. By doing all those leisurely, enjoyable things in the middle of the night, I was rewarding my brain for waking me. They advised me that if I didn't want to continue to wake up every night, the best thing to do the next time I woke was to get up and do something that I hated doing, like cleaning the oven or washing a floor. I tried it. I got up and cleaned the oven. Apart from the odd occasion, I can honestly say I've not unintentionally woken up on a regular basis since.*

**– Adrian**

Night owls beware! White matter connecting brain areas that generate sadness and depression has been found to be in worse condition than in early-bird brains.

## Chapter takeaways

- Stress is a friend so long as it doesn't overstay its welcome.

- Be the driver not a passenger of the *Stress Express*. Stop when you need to stop.

- Rest is vital if you want to give your brain a chance to catch up with much needed repair work.

- Keep the last sixty minutes before going to bed as stress free as possible. Don't *try* to get to sleep and if you do keep waking up, go and do something that you hate doing!

- If you want to reduce cortisol and stress levels – make space for GOM Time.

- When you're next up for an award remember to thank cortisol.

# "Smart" Drugs

## Everyone's on drugs!

Do you numb the pain of a headache with aspirin, paracetamol or ibuprofen? When you have a sore throat, do you pop into a shop and buy throat lozenges? Do you like to give yourself a bit of a boost each day with a cup of coffee or maybe you prefer tea as your morning kick start? Perhaps you like to unwind in the evening with a glass or two of alcohol? Possibly you enjoy smoking tobacco or even cannabis?

Whilst the last of these is broadly considered a drug due to its (il-) legal status in many countries, it is no more or less a drug from the perspective of science than all the aforementioned chemical substances that many of us routinely take to improve the way we feel or perform.

The whole world is at it. The reason this can be said with such confidence is because of the broad dictionary definition of a drug:

*"A chemical substance used in the treatment, cure, prevention, or diagnosis of disease or used to otherwise enhance physical or mental wellbeing."*

This definition goes way beyond an antibiotic, an anti-inflammatory or an anti-depressant prescribed by a doctor, not to mention a whole plethora of remedies available over the counter at chemists. Pretty much anything you put in your mouth or inhale, that changes the way you feel, fits the bill.

Our history of eating, brewing and smoking leaves, shoots, roots, berries, buds and herbal mixtures with a view to making life more bearable or pleasurable pre-dates the discovery of the wheel by several millennia. This may be, but what's becoming a real concern these days is the level of drugs that *healthy* people are taking to enhance their wellbeing, in particular nootropic compounds. Usually referred to in the media as "smart drugs," they are known to increase alertness, energy levels and improve various cognitive abilities.

Numerous new chemical substances have popped up in the past couple of decades, which enable already fit and able people to achieve more – both physically and mentally. The considerable pressures to study, work and play harder than ever before, just to keep up with the breakneck pace of life in the 21st century, are driving more and more people to give taking them serious consideration. Anyone tempted by the lure of these "chemical assistants" should be aware of and fully understand all the pros and cons.

## On top of the world

The list of chemicals routinely self-administered with the express goal of improving physical capabilities is vast. We have all heard about controversial track and cycling stars who cheat by illegally dosing themselves with performance enhancing drugs to increase strength and stamina; anabolic steroids to increase muscle mass probably being the first to spring to mind.

We know from the previous chapter how another hormone – cortisol – works, so you'll not be surprised to hear that anabolic steroid hormones can also have a profound effect not just on the body but also the mind. "Roid rage" is a term associated with people who overdo steroids. As you have no doubt guessed, these people have a pretty short fuse. Taking anything as powerful as a hormone without medical supervision is always likely to end in tears. This is one of the reasons why using anabolic steroids without a prescription is illegal. Under different circumstances i.e. when prescribed for people suffering from conditions that cause muscle wasting, exactly the same substance is perfectly legal.

Cocaine is an illicit recreational drug notorious for its use by musicians, models and film stars to fuel wild parties and all-night drinking binges. Yet the humble coca leaf, the raw material that is highly processed and concentrated to produce the infamous white powder, has been quietly consumed for hundreds of years by cultures living high in the Andes and with a very different goal in mind. A single leaf can be brewed for consumption as a tea (known locally as maté de coca) or simply chewed to extract the juices.

 If your skull was used as a drinking vessel it could hold about three pints.

The use of coca leaf in the Incan culture was once reserved exclusively for religious ceremonies and by their royal family. Before long Incans of all classes were allowed to chew it in order to enjoy its mildly euphoric, appetite-suppressing and stamina bolstering effects. Indeed, it has been argued that the incredible high altitude architectural feats of the Inca nation, Machu Picchu being the crowning example, would not have been possible without the effects of the humble coca leaf.

> For decades recreational drugs like magic mush-rooms, ketamine and ecstasy were thought to *trigger* mental health problems. Now they are being inves-tigated for their potential to *treat* obsessive compulsive dis-order, depression and post-traumatic stress disorder.

When it was first brought back from the new world to Western Europe it was purified into cocaine extract and prescribed by doctors for a variety of ailments. It was even championed by Sigmund Freud, Queen Victoria and the Pope, no less! Before long though the negative aspects of the highly concentrated form of coca reared their ugly heads (addiction, heart problems, psychosis etc.) and, as quickly as it had become popular, it was swiftly outlawed.

In its unconcentrated form it is still commonly and legally used throughout much of South America. Yet in its highly purified and concentrated forms, as well as heart attacks and psychosis, crack and cocaine causes mind-bending chaos. It hijacks the habit-forming areas of the brain, resulting in addiction and all the crimi-nal behaviour that usually tags along to help fund it. The point here is that a drug in its less potent natural form can be enjoyed and exploited for its ergogenic (work promoting) properties without causing major problems, but once its potency is dramati-cally increased the same drug can wreak havoc in a person's life and in society as a whole.

# My cup of tea

*After successfully finishing my PhD and completing a lecture tour presenting my brain-scanning experiments at universities across North America, I spent a few months travelling around South America.*

*As Sod's Law would have it, I arrived in Peru during the month of the year when they close the Inca Trail for cleaning. Undeterred I travelled to Aguas Calientes by train so that I could walk up the mountain to see Machu Picchu; a lifetime ambition of mine. Having travelled quickly from low to high altitude, the following morning I woke up feeling dizzy and nauseous. Despite not really thinking I could stomach it, I went to a café on the main street for breakfast. Upon explaining my need for plain toast with nothing on it to the friendly waiter, he refused my order for coffee or tea, insisting that I take maté de coca instead.*

*Having spent much of my teens and early twenties partying all over London, I had passed up the opportunity to try cocaine on numerous occasions. I reasoned that if I never tried it then I would forever find it easy to refuse, having no personal experience of its effects. Despite my reluctance, the waiter insisted that a cup of coca tea was completely incomparable to cocaine and that it would instantly cure me of all my symptoms. Eventually I succumbed. He was absolutely right. The dizziness and nausea instantly evaporated to be replaced with a mildly euphoric state of mind. Needless to say that for the rest of my trip whenever maté de coca was offered at breakfast as an alternative to coffee or tea I shunned my usual cuppa every time. I have still never tried cocaine and intend to keep it that way!*

**– Jack**

# Not so smart drugs in schools

A new phenomenon of drug abuse is sweeping our schools and universities, and it's not just hedonistic youngsters sneaking around looking for a quiet place for a crafty cigarette or spending weekends popping pills in nightclubs and at warehouse parties. The high achievers are at it too and they're not after cannabis, ecstasy, cocaine or magic mushrooms. They want to get their hands on the so-called "smart drugs." Surveys have revealed that the two top motivations for taking smart drugs have nothing to do with getting high and much more to do with getting higher grades. They are:

1. Gaining the competitive edge over peers.

2. Making mundane tasks seem less boring and thereby increasing productivity.

Studious students are buying methylphenidate (aka Ritalin) from fellow pupils diagnosed with ADHD (Attention Deficit Hyperactivity Disorder) for whom the drug is prescribed to help their brain to focus on the task in hand and sustain concentration. Highly inflated prices are paid on the scholastic black market in the hope of being able to study harder and convert any stray B's into A's. When "everyone else is doing it" there is inevitably a perceived pressure to follow suit so as not to miss out.

That many kids are finding themselves compelled to illegally acquire drugs that might help them to outperform their peers in school is extremely disturbing for several reasons. It is worrying that healthy children are putting prescription-only pharmaceuticals into their systems without consideration for whether or not it is safe to do so. Aside from a whole range of potentially dangerous interactions with other medications and the high likelihood of incorrect dosage, one huge concern is that during adolescent years, when young brains are undergoing a process of considerable remodelling, the presence of methylphenidate in their system might cause problems with this vitally important process.

The other big concern is that upping the normal dose leads to a mild sense of euphoria, via interactions with the dopamine-driven pleasure pathways of the brain (Reward Line), which means that methylphenidate can be highly addictive. Rest assured: when taken to treat ADHD as prescribed by a doctor methylphenidate (aka Ritalin) is NOT addictive.

It is argued that, when the same drug is prescribed for ADHD sufferers, the benefits gained by enhancing their ability to focus and concentrate for prolonged periods of time significantly outweigh the potential negative consequences. Indeed, untreated ADHD is often associated with poor educational experience and a high dropout rate, which sadly prevents many of these children from reaching their true potential; and so treatment is deemed worthwhile.

There are differences in how methylphenidate affects the brains of healthy people versus those who have a brain disorder. When, for example, methylphenidate is used to treat ADHD, improved sustained attention is not the only benefit: impulsiveness and love of risky behaviour tangibly decreases. On the other hand, when

## Good drug – wrong brain

In an experiment to investigate risky decision making, 40 ADHD-free women were dosed up with methylphenidate. All made wildly risky choices, the financial consequences of which proved to be disastrous!

A bucket of water may put out a wood fire, but using the same bucket of water on a chip pan fire could be absolutely catastrophic.

taken by anyone who doesn't have the symptoms of ADHD, quite the opposite effect can be induced.

## Whistle while you work

Experiments have looked into the effect of methylphenidate on healthy adults when being used for non-medical and purely self-enhancement purposes, e.g. improving productivity in the workplace. When asked about its appeal, participants said they feel more alert, attentive, energetic, even if there was very little in the way of objective data to support this impression. It also appears to have a mild positive effect on both spatial visual working memory (holding in mind the relative positions of several objects) and planning abilities that are so important for making good rational decisions. However, one major drawback is that, despite helping to increase productivity when it comes to routine work, there is some evidence to suggest that it suppresses creative thinking.

It is not just kids dabbling in the black market for methylphenidate in pursuit of squeezing more work hours out of themselves, it seems that their parents are at it too! In a poll of 1400 readers of the world famous and highly prestigious scientific journal, *Nature*, 20% admitted to taking smart drugs to improve their work rate. Around half of these had used methylphenidate and others had used a drug called modafinil. Modafinil was approved by the USA's Federal Drug Agency in 1998 for the treatment of narcolepsy – a condition where people cannot help falling asleep at the drop of a hat throughout the day. Since then it has become extremely popular as a smart drug in both the civilian and military workplaces – used to to help keep people alert when jetlagged or sleep-deprived.

Modafinil has been tested extensively in people free of medical conditions. Studies done within the military showed that staff who had been kept awake for 40 consecutive hours felt much more alert and able to function effectively having taken modafinil. In another study a single dose of modafinil was given to medical doctors after a long shift. They felt that they worked more efficiently with a notably enhanced ability to focus attention. Their decision making was also proved to be better thought out and less impulsive.

Modafinil seems to have a positive impact on decision making when tired by encouraging users to consider the options more carefully. For this reason it has been employed to help those with pathological gambling problems. Gambling seems less attractive whilst under its influence; and if people do gamble anyway it reduces their soft spot for the riskier decisions.

Positive impact on decision making aside, it seems that the main allure of modafinil is that it helps people to keep going under duress and fatigue, as well as improving attention and memory. All very attractive and particularly relevant in the modern professional climate where the work–life balance for many feels increasingly like it's slowly tipping more towards work.

## Rock the Basal Ganglia

There is a great drug-free option available to everyone to boost physical stamina – music. Considerable evidence suggests that high tempo, happy-sounding tunes that are to your own personal taste can increase strength, speed and stamina, whether you are running, cycling, doing circuit training or swimming. And, despite all the extra work, your perception of exertion is much lower!

Activity in your Basal Ganglia – the brain area responsible for initiating movements – increases in response to the beat of the music.

## Nootropic drugs – big dilemmas, grave outcomes

The use of smart drugs has been described as cosmetic psycho-pharmacology and cognitive enhancement, amongst other things. Call it what you will, the use of pharmaceutical agents like methylphenidate (Ritalin) and modafinil (Provigil) for performance enhancement and sleep avoidance is skyrocketing.

New, interesting moral dilemmas have cropped up in recent times. Some doctors have said they feel ethically obliged take some form of smart drug when their punishing schedule leaves them feeling fatigued and thus compromised in their ability to make critical life and death decisions. They feel a moral duty to reduce the errors that inevitably creep in due to sleep deprivation by taking drugs like modafinil to improve concentration and alertness.

Doctors in the USA also complain that they are coming under increasing pressure to dish out prescriptions for smart drugs to help healthy patients keep up with the incessant demands of increasingly competitive working, academic or social environments.

These dilemmas may both sound quite straightforward at first glance but the upshot is that, in either case, it is quite simply illegal for these drugs to be prescribed for such purposes. Irrespective of issues of legality, it exposes all concerned to some potentially dangerous health issues and opens up a whole can of complex social consequences.

In sleep deprived individuals a single dose of modafinil does have a strong positive effect on executive function and improvement in memory, an effect that wears off during continued sleep deprivation. Were they to take a single dose when not sleep deprived, they would find it has the opposite effect – it actually induces drowsiness! Furthermore, repeated doses of modafinil when not sleep deprived increases both the positive and negative emotional state, which means you simultaneously feel slightly happier *and* more anxious.

In the overall timescale that we humans have been dabbling with mind-altering substances, these particular drugs really haven't been around very long. We cannot possibly know what the negative consequences of using these smart drugs over many years might be. However there are other compounds out there that our ancestors have consumed for centuries and so the long-term outcomes are well characterized. As a direct consequence we would argue a person is on safer ground using these old favourites rather than the new kids on the block.

# The original smart drug

There is another stimulant drug that raises alertness by increasing the brain's ability to produce energy and by inducing the release of increased quantities of a variety of brain messenger chemicals. The speed at which information is absorbed from the outside world is accelerated and motor processes are sharpened leading to quicker reflexes.

Similarly to methylphenidate and modafinil this wonder drug doesn't really make you more intelligent *per se*, it just helps you carry out routine behaviours more efficiently. Beyond these immediate effects it turns out that this miraculous drug even protects against several illnesses strongly associated with the aging process in body and brain. Moderate daily doses protects against Type II diabetes, liver disease, Alzheimer's and even Parkinson's disease.

So, what is this wonder drug we are talking about? A compound that not only increases brain efficiency in the short term, but also protects against several horrendous medical conditions in the long run?

Well, ladies and gentlemen, get the bunting out and cue a drum roll – the name of this smart drug is . . . caffeine. That's right, common-or-garden caffeine.

Caffeine's chemical structure is almost identical to a messenger substance called adenosine. When it's released into the gap between one brain wire and the next, its job is to make the second brain wire *less* likely to fire electrical messages. For this reason, it's known as an inhibitory neurotransmitter. As caffeine is almost exactly the same shape as adenosine it can slot perfectly into the special receptor in the wall of the second brain wire perfectly – but without actually causing the receptor to switch on and do its job of slowing things down.

Caffeine is a receptor blocker. It clogs up receptors so that the adenosine can't get in to do its work of suppressing electrical messages and, as a consequence, disinhibition takes place. Blocking the inhibitory effects of adenosine makes brain cells *more* active by, in effect, taking the chemical hand brake *off*. This has a direct impact on many different messaging chemicals that bridge the gap between one brain wire and another, including dopamine, glutamate, noradrenaline and acetyl choline to name but a few.

## Full of beans

Caffeine increases levels of the excitatory neurotransmitter glutamate on the Reward Line accounting for the pleasure people derive from drinking coffee or tea. Like all good things though, we can have too much of it. Overdoing caffeine means inducing so much disinhibition in your Basal Ganglia – the brain area that initiates movements – you end up getting the shakes.

## Varying effects

Lifestyle factors can have a powerful impact on how caffeine is metabolized and thus its effects on an individual's brain function. Smoking tobacco doubles the rate at which caffeine is removed from your system and so the effects of coffee are shorter lived in smokers than non-smokers.

Conversely, in women using the contraceptive pill, caffeine remains in the system for twice as long as normal, so each dose of caffeine has a longer-lasting effect and so successive doses build up to higher concentrations.

Given how long modern day smart drugs have been on the scene and the insufficient long-term safety data, if you are in need of a bit of a buzz, it's probably best to stick with a smart drug that has a proven track record. This is particularly the case for adolescents where the risks of undesirable side effects are even greater due to the massive overhaul their brains go through during the teenage years. It would be wise where possible to avoid any disruption to the course of these natural changes.

Caffeine has been around for a very long time and with 50% of the world's population regularly drinking coffee, the risks involved are well understood. As with all legal drugs, as long as you are careful to keep your daily dose within the low to moderate range, the risks are minimal.

If taken in moderation, there are also often significant gains. A midlife intake of three to five cups of coffee per day appears to substantially reduce the risk of developing Alzheimer's disease. In later life elderly people who drink up to three cups of coffee per day exhibit significantly delayed cognitive decline than those who abstain from the magic bean. As we've heard, its neuroprotective qualities don't stop there – it also protects against Parkinson's disease.

## Brewing for survival

We humans co-evolved with a variety of plants and fungi that contained chemical substances that, through some mechanism or another, enhanced our physical or mental wellbeing. Our ancestors may well have been introduced to alcohol when thirst or hunger drove them to risk biting into rotting food where yeast spores had been hard at work on the sugars found within – apples, honey, pears, grapes or cereals. If you've ever seen wasps getting so drunk on rotting fruit that they can't fly straight, you'll be able to imagine how common accidental drunkenness would have been in ancient times.

The desire to repeat the feelings of physical and mental wellbeing that result from mild intoxication with alcohol would have provided the motivation to seek the experience out on a more regular basis. Before you know it, we'd figured out how to ferment these juices intentionally, leading to the wide variety of the ciders, meads, wines, spirits and beers, all now readily available whenever we wish to put ourselves in a different frame of mind.

As a poison, alcohol was able to kill off the various pathogens found in the water supply that might have otherwise led to sickness. Early peoples with a penchant for alcoholic beverages would have had a survival advantage over their peers. The alcohol in their drinking water made it safe to drink and this helped our ancestors to survive for long enough to reproduce.

Alcohol is not just poisonous for many bacteria and parasites, if we couldn't break it down it would poison us too. Fortunately, some of our ancestors happened, completely by chance, to possess a liver enzyme capable of breaking down the toxic alcohol – alcohol dehydrogenase. Those with this enzyme could enjoy the benefits of disease-free water without succumbing to the damage done by the alcohol itself. The ever-presence of alcohol in the diets of many Caucasian races thus favoured the survival of those whose livers could produce this enzyme. However, other industrious races came up with a different solution to making water safe to drink.

Elsewhere in the world, in Africa, Asia and the Middle East, the solution to making water safe to drink was to boil it. As boiled water tastes pretty horrible the addition of leaves or berries to the brew became commonplace. And thus the habits and rituals of tea and coffee drinking were born.

The safe drinking of toxic alcohol requires the gene to produce alcohol dehydrogenase, which most people of Caucasian origin possess, but many of Asian origin lack. As the chemical substances in tea and coffee are not poisonous, anyone can enjoy them regardless of their genetic makeup.

## What is a hangover?

Your kidneys constantly squeeze all the juices out of the blood as it is forced through millions of tightly-coiled blood vessels. Everything your body does not need, like the waste products of metabolism, are passed on to the bladder so that you can be rid of these in your urine. Water, on the other hand, is selectively reabsorbed back into the bloodstream across special channels that can be opened and closed to ensure you stay adequately hydrated. Unless, that is, you have alcohol in your bloodstream.

Alcohol paralyzes the special channels that reabsorb the water back into the blood so it all travels down to the bladder, which is why you urinate more when inebriated. Consequently, you become increasingly dehydrated the more booze you drink.

Your brain itself has no pain receptors; it doesn't feel pain. There are, however, pain receptors in the three-layered sack in which your brain resides – known as the meninges. When you are dehydrated these layers all press together, stretching the pain receptors within. This is where the head pain associated with the dreaded hangover comes from. Rehydration is the best cure for a hangover but it's not always as straight-forward as just drinking water. If you've been sweating a lot then you will have shed a lot of salts in your perspiration. Your kidneys can't work properly if they are salt depleted which is why people often crave salty foods when drunk. A glass of rehydration salts before bedtime can be just the ticket to help you feel better in the morning by helping your kidneys to reabsorb water.

## Wear and tear

Why does caffeine seem to protect the brain? Nobody really knows for sure. It may be that by blocking adenosine's inhibitory effect the resulting acetyl choline boost helps to keep Alzheimer's disease at bay. Similarly, the surge in dopamine levels may be key to keeping Parkinson's disease away. Or it might just be the fact that coffee and tea, especially green tea, are jam packed with antioxidants that help reduce general, age-related wear and tear on brain cells.

A good night's sleep is much more important than anything else when it comes to healthy aging. If you have trouble sleeping then reducing caffeine intake and limiting it to mornings would be a good first step. But if sleeping's not a problem, then drinking a few caffeinated beverages a day might just be the tonic for getting more done and enjoying a healthy brain well into old age.

The longer a drug has been around the more we will know about its long and short-term effects. The newer smart drugs like methylphenidate (Ritalin) and modafinil (Provigil) are becoming increasingly readily available. Their use in schools, workplaces (and before long probably retirement homes too) is on the rise.

We may not know for certain all the outcomes of regularly taking these substances but what we do know is that taking any prescribed drug that hasn't been officially prescribed always has the potential to lead to dangerous, unnecessary health risks. We also know about the high likelihood of damaging interactions with other drugs if smart drugs are taken without medical advice alongside already prescribed medicines.

The upshot is, if you do feel the need to rely upon chemical assistants to help you keep up with your busy life, if the substance you're considering hasn't been around for a very long time and hasn't been prescribed to you, you're always going to be playing "Neuro Roulette" – no matter what anyone tells you.

## Chapter takeaways

- Got a hangover? As well as rehydrating, if you've been sweating a lot, make sure you replenish your kidney's supply of salts.

- It is man's tampering that has given the coca leaf such a bad name. In its natural form and in moderate quantities, it's fairly harmless. The concentrated forms on the other hand are dangerously addictive.

- The latest smart drugs may be great for people with pre-scribed needs, but for those just looking to up their perform-ance they could well prove to be *not* such a smart choice in the long run.

- If you are in need of a little brain optimization and want something that's legal and has a proven track record, caffeine is your safest bet.

- Tea and coffee might not be to your taste. If not, get your Basal Ganglia jumping with whatever happens to be your current favourite high-tempo beat!

- Thank your ancestors' ingenious use of alcohol or caffeine for you being here today.

# Hold On to Your Marbles

## Over the hill

As far as the *structure* of your grey matter is concerned, once you're past your mid-twenties, it's downhill all the way. Your brain actually begins to shrink. But that doesn't mean that brain *function* peaks in its mid-twenties. In many regards several cognitive abilities actually become more efficient as you age, despite the structural degradation. In this regard, a better system of wiring is more important than having more wires.

Throughout middle age and beyond, the branching antennae (dendrites) that pick up signals from other neuronal brain wires progressively retract and lose the spiny tendrils that are studded with synaptic connections. The exception to this rule being the neural pathways involved in mental activities that you regularly challenge your brain to tackle. This helps to ensure that the relevant synaptic connections are maintained. Unfortunately, whatever you do, the myelin wrapper around the wires that speed up electrical transmissions will gradually lose its integrity, this part of the process starts to kick in during your forties.

These processes all lead to a thinning of your cortex – the outer layer of your brain within which billions of wires share their electrical information across your ever diminishing numbers of synaptic connections. As a consequence of all this brain shrinkage, the valleys at the surface of your brain get wider, the peaks get narrower and deep inside, the fluid-filled spaces (your ventricles) become larger. And, to add insult to injury, your neurotransmitter systems, the chemical messengers that cross the gap (synapse) between the end of one brain wire and the beginning of the next, become incrementally less efficient as each day goes by.

This gradual decline is inevitable and an inescapable fact of life. If we all lived to be 150 years old, we'd all display obvious signs of Age Related Cognitive Decline (ARCD) as a result of these natural processes of brain aging. Being around that long would mean that at some point we would all have succumbed to the forgetfulness, distractedness and compromised problem solving that characterizes ARCD.

Sadly, for many people ARCD kicks in far too early on in life. Whilst some people still function just fine well into their nineties, only starting to show signs of decline as centenarians, for others it can begin to interfere in the tasks of daily life much sooner – perhaps even in their forties or fifties.

We've all encountered it. Whether it is someone who constantly forgets that they've already told you their favourite story many times before, a person getting lost on a journey they have completed on numerous occasions, or an elderly relative who always calls you by the wrong name.

Witnessing cognitive decline in someone else can be one of life's most harrowing trials, but for the person actually experiencing it, well that's something else altogether. The transition from strength and vigour to weakness and frailty is tough enough to deal with, but lacking the capacity to care for your own needs, not to mention the loss of dignity associated with losing your independence, can be devastating.

ARCD may be an inevitable thing in the long run but there is, you'll be pleased to hear, some good news. The speed at which its processes unfold varies greatly from person to person. In many cases, inherited genetic conditions notwithstanding, it seems that these differences are largely to do with factors over which you can have significant control. Rather than abandoning yourself to fate, you can instead establish what you can do to hold onto your marbles for as long as possible – and then do something about it.

The really great news is that many of the things you should be doing to ensure you maintain a healthy brain for as long as possible are enjoyable activities. Activities that get you out and about, that involve interacting with other people and generally having a good time.

# Putting the brakes on

Thanks to the incredible advances in medicine, life expectancy continues to lengthen. It's great to know that we're all likely to be around a lot longer, but being around longer also means that our chances of experiencing ARCD are also increased. This leads to the billion dollar question: what can each of us do to hang on to our marbles for as long as possible, right up until the day we do finally conk out?

Well, there are several things that we can all start doing right now. Many of them are simple things that we've already covered in this book, which unless you've skipped straight to this chapter you are already aware of. Out of all the practical bits of advice, there are two highly effective brakes that you should seriously consider applying sooner rather than later, if you haven't already done so, to slow your descent into ARCD.

## 1. Reduce free radicals

Without drilling down into too much detail a free radical is like a bull in a china shop. If not brought under control, the bull will eventually smash the shop to smithereens. A free radical is to the structure and fabric of your brain, what that bull is to the china shop.

As with all classic tales the story of free radicals involves a lifelong struggle between good and evil. The goodies are the antioxidant foods which, when regularly introduced into the battleground within your body and brain, willingly give up a spare electron to any free radicals that happen to be knocking around. This renders them harmless before any damage can be done. In the absence of antioxidants, the desired electron will end up being snatched indiscriminately from a nearby piece of cellular machinery, introducing faults and problems that accumulate over a lifetime to interfere with normal brain function.

The heroes in this particular saga are the fresh fruit and vegetables that provide a plentiful supply of brain-tissue-protecting antioxidants whenever they are ingested in sufficient quantities. The baddies are the substances that are routinely taken into our bodies through our lungs and our guts that are the source of free radicals in the first place. The villains in question include fatty foods, exhaust fumes and cigarette smoke.

## 2. Keep your blood pipes clear

The best three possible things you can do to keep your blood pipes clear are: stay away from smoking, keep your intake of saturated fat to a minimum and exercise with at least moderate intensity every other day.

The typical Western diet is abundant with saturated fats. Without wanting to make this sound like some sort of conspiracy theory, adding saturated fats to processed foods helps big multinational companies to sell more food and boost their profits. Processed foods are often laced with saturated fats because quite frankly, as well as increasing shelf time, it's the best way to make rubbish food palatable.

Make time to prepare fresh food, eat chicken rather than ham and go for turkey over beef. It doesn't mean you should never again eat the foods you love; there certainly wouldn't be much fun in that. It's all about eating the lean stuff daily and the delicious fatty stuff infrequently, as a treat. The less often you eat it – the better it will taste!

Animal fats and the tars in cigarette smoke clog up your arteries, blood vessels that must constantly supply oxygen and glucose to every cell in the brain, every moment of every day. This process (known as atherosclerosis) causes sticky plaques to accumulate along the inside of your arteries gradually clogging them up and making it increasingly difficult for blood to pass through. But that's not all. Not only do atherosclerotic plaques make these vital blood pipes narrower but it also makes them more rigid. Usually, as each heart beat forces more blood into your arteries, they are stretched wide and then the elastic walls snap back into place (what you feel when take your pulse) squeezing the blood ever onwards before the next surge of pressure is produced by your beating heart. Atherosclerosis makes arteries lose their elasticity, making blood vessels even more inefficient in piping much needed blood.

The reduction in the amount of blood that can be carried at any given time due to narrowing of vessels, along with the loss of elasticity preventing it from being effectively shunted along, has a negligible effect from one day to the next. Yet the incremental reduction in effective transport of that precious oxygen payload gradually takes its toll over many years. You are unlikely to notice the impact from year to year even, but from decade to decade its effects can become fatal. If the supply of oxygen-rich blood to the heart is completely cut off, even if it's just for a few minutes, parts of your heart muscle will stop working and soon begin to die off completely – never again to function properly.

The first sign of trouble usually comes after many years of gradual changes in the blood vessels supplying your heart with everything it needs to do the energy-consuming job of keeping its muscles permanently pumping. By the time someone has a heart attack many of their blood vessels will have narrowed to three quarters, possibly a half, and sometimes even a quarter of their original widths! Although heart attacks are triggered when the blood supply to part of the muscle tissue is completely cut off, the heart can also weaken over long periods of time due to the gradual reduction of blood supply getting through to it. And a weakened heart means that oxygenated blood is less efficiently delivered to your brain.

# Brain attack

Although heart attacks are pretty common and fairly well under-stood, most people seem to be unaware that the same process can also take place in the brain. This is why a high dietary intake of saturated fats is associated not only with a high incidence of heart failure, but also with a high incidence of stroke. If a heart attack is the name for the situation whereby the blood supply to a certain chunk of heart tissue is cut off and the tissue stops func-tioning properly, then a stroke might be better described as a "brain attack" because it is exactly the same problem but with a different organ.

A body of evidence is accumulating to suggest that the symptoms of cognitive decline are caused by a series of tiny undetected strokes that occur throughout life. "Silent Cerebral Infarcts" as these mini-strokes are known, may also play a role in causing migraines when they damage the white matter that ferries electrical information from one part of the brain to another. Our only hope of protecting ourselves against this is healthy blood flow.

## A most misleading word

When used in its medical context there could not possibly be a more misleading word than the word "stroke." It conjures up images of something that is gentle, soft, mild and perhaps even soothing. If these four words do spring to mind, think of four words that are the exact opposite. Whatever words you have thought of, you will now have a far more accurate description of just what a "stroke" is all about.

If the threat of a heart attack doesn't make you give that double sausage, double egg, chips and fried bread breakfast feast a miss, then maybe the equally likely threat of a brain attack will make you think twice. Again, it doesn't mean you should completely cut out eating your favourite fry up but if you can, try to make a habit of eating foods low in saturated fat, with the occasional celebratory gastronomic blow out now and then.

## Brain change activity

Avoiding saturated fats and toxic smoke to keep the blood vessels wide open and flexible is only half the story. The reason that regular exercise is so good for your brain as well as your health in general is that it strengthens your heart. Eating healthy food (fruits, vegetables, fish, nuts, herbs, spices, pulses, lentils and wholegrain cereals) ensures that your blood is filled with nutrients. If the very pump that distributes this nutrient rich broth to every cell in the body is strong then it will reach every nook and cranny of your brain. Your brain is constantly in need of raw nutrient materials to build new connections, create new neurotransmitters, carry out maintenance on its wires and keep the myelin go-faster wrappers of those wires in a state of good repair.

If your heart is allowed to fall into disrepair by, for example, driving instead of walking or by never getting around to spending half an hour getting out of breath every other day then it will become weaker. So when it comes to keeping every part of body and brain fully stocked with all their basic requirements, it will fall short, it will fail to deliver. It's as simple as that.

It's never too late, but the older you get the more vigilant you need to be in terms of eating healthily and exercising regularly. Decades of self-abuse can be turned around and, indeed, the weaker your heart has become the greater the benefits you will perceive once you start making the lifestyle changes necessary to strengthen it. You'll have more energy and you'll feel more motivated to get out there and do something that's active, sociable and hopefully mentally stimulating. The result being that once you are out there doing it, you'll find that your mood will naturally improve, you'll be inspired to do it more and you'll be more open to trying out new activities.

## Cog-turning activities

**Bilingual.** Not just being bilingual, but regularly having conversations in both languages can also really boost brain powers. It's been discovered that compared with a purely monolingual brain, bilingual brains have better task switching and sustained attention capabilities that hold you in good stead well into old age.

**Table Tennis.** Not only does it get your heart pumping, it's also great for keeping your grey matter active. Your brain is fully engaged as hand–eye co-ordination, reaction times, spatial awareness and having to think up new strategies on the hop are all brought into play.

**Juggling.** Juggling helps induce tangible neuroplasticity. With its responsibility for monitoring, quickly reacting to and guiding movements towards fast-moving objects in peripheral vision, it really puts your Intraparietal Sulcus (IPS) stop through its paces. And, on top of all this, it pushes your Visual Lines to their limits. An all-round, top cog-turning activity.

This new found drive to take part in a range of different activities will, if you participate regularly enough, stimulate reinforcement of the connections between all the different brain areas involved in carrying out those activities. And, if they are sufficiently stimulating, they will even help you build something called cognitive reserve that can keep dementia at bay for longer. Physical activity is the key to a healthy heart – which is, quite literally, the driving force of mentally stimulating, sociable activities that challenge your brain to change.

## Building cognitive reserve

The Einstein Aging Study (so named because it was conducted by the Albert Einstein College of Medicine – Albert himself was not involved!) followed 2000 people aged 70 and above who were residents of the Bronx district of New York City for four years. Every year these residents were put through a variety of tests to monitor changes in their physical strength, balance and coordination, along with a wide variety of cognitive abilities. As well as undergoing these tests, images of their brains were captured with an MRI scanner.

The aim was to examine the impact of specific lifestyle choices on the various factors that influence a person's ability to look after themselves, with ARCD high on the list of priorities. Specifically, they were interested to know if there were any particular hobbies, games or social activities that might somehow help to condition the brain to resist the cognitive impact of the metabolic inefficiencies that inevitably crop up as we age.

They found that four activities were associated with a significantly reduced likelihood of developing the symptoms of cognitive decline: playing a musical instrument, playing chess, dancing and reading all seemed to have a positive impact on slowing the rate of cognitive decline. It was noted that none of these activities made the slightest difference to the outcome unless they were practised regularly.

They also found that those who were often socially engaged, who took regular moderate to intense exercise and that participated in the above activities more than once a week enjoyed a much longer period of dementia-free life.

You'll no doubt recall how we previously talked about your brain physically changing to better accommodate any specific behaviour that is practised intensely, regularly and consistently. It should come as no surprise to you that now and then, occasional engagement with the above activities does little to keep dementia at bay.

What's more, all of the above activities are mentally taxing – the other defining feature of activities that actually inspire the brain to make changes. If you don't up the *ante* in terms of tackling more and more challenging versions of the same activity then the brain will stop making the necessary changes for further improvements. Let's take a closer look at each of the activities:

*Playing a musical instrument* involves manipulations of an object with various body parts (e.g. fingers, lungs and mouth for most wind instruments) to produce tightly coordinated and rhythmically precise sounds. These sounds must match the desired musical notes that are usually effortlessly translated from visually scanned marks on a piece of paper, denoting the pitch, onset, duration and style of each note. All whilst listening keenly to sounds produced both by oneself to ensure accurate rendition but also in light of others to constantly establish that the desired overall effect is being created.

As you can imagine, all of these processes are extremely cogni-
tively demanding to one degree or another. Hence playing a
musical instrument induces changes that enable brain areas in the
prefrontal cortex (PFC) to exert better control, through tighter
integration, over the many other brain areas that all need to be
carefully coordinated to successfully hit the right notes at exactly
the right time. It is this dense connectivity between PFC brain
areas, that seems to confer the miraculous marble-preserving
facility of auxiliary brain networks that take over mental functions
that would be otherwise lost when important brain areas are irre-
versibly damaged by the aging process.

*Chess* requires potential moves of both players to be imagined
and held in mind so that further moves can be thought through
and evaluated. Opportunities and pitfalls of each potential
sequence of moves must be analyzed to select the best strategy.
The more moves in advance a person tries to plan, the harder the
brain areas in their Frontal and Parietal Lobes that support working
memory (the Hold Line) are pushed, to try to keep in mind where
all the pieces would stand after each imagined move.

The harder working memory is put to the test during the day, the
more work will be done overnight to reinforce the synapses connecting Frontal and Parietal brain areas to increase its capacity for
next time. As we know from a previous chapter, boosting working
memory is a very powerful way of increasing IQ as it provides a
sturdy foundation upon which problems can be better solved.

*Dancing* is a quintessentially social event. Maintaining regular
social contact with other people is known to have a powerful positive impact on mental health and wellbeing by making individuals
feel connected to their community. Most dances bring two humans
into direct physical contact, which triggers the release of oxytocin
from their Pituitary stop into the bloodstream. This is a neurohor-

mone that induces feelings of trust, comfort and a sense of belonging, again increasing feelings of social connectedness and wellbeing.

The cognitive challenge presented by dancing is no easy one. Your Auditory Line has to create the perception of music and find the beat. Your Visual Lines have to make sense of the dance moves demonstrated by your instructor. The Mirror stop is involved in both observing the actions of others and performing those actions yourself – it is critical to the process of learning skills from others. Your Basal Ganglia stop initiates body movements and must do so in time to the beat of the music. And your Cerebellum has to constantly fine-tune the signals sent to your muscles to maintain balance. All of this must be coordinated with the movement of your dance partner. As well as keeping you light on your feet it also helps to sustain an agile mind. A better whole brain workout you'd be hard pushed to find.

*Reading* involves converting strings of letters into words, words into sentences and keeping the meaning of one sentence in mind in light of those preceding and following it. Images must be conjured up in the mind's eye, sounds in the mind's ear, tastes and smells in the mind's mouth and nose. Previous chapters of whatever book you happen to be reading must be brought back to mind in order to interpret new events and imagine future scenarios in anticipation of the most likely outcomes. The trajectory of the plot must be tracked whilst a dynamic impression of the characters' personalities is assembled, not to mention the inter-relationships between them and knowledge of what they do and do not know during the intertwining narratives. Fabulous mental gymnastics are performed any time we lose ourselves in a good novel.

## Nuns on the run

In 1986 the USA's National Institute on Aging funded a study in which 678 Roman Catholic nuns were tracked over many years in an effort to better understand Alzheimer's disease. They were tested on a variety of cognitive abilities, including holding a list of words in memory, how many animals they could name in 60 seconds and coin counting. The idea of studying nuns was inspired by the fact that their life experiences tend to be very similar. So any differences in the onset of age-related psychiatric problems could be accurately attributed to nature (their genes) as opposed to nurture (their life experience).

To this end, once the nuns died their brains were donated to science so that they could be examined for the telltale signs of Alzheimer's disease. Unsurprisingly many of the sisters who had developed symptoms of Alzheimer's disease exhibited a large amount of Beta-amyloid plaques and neurofibrillary tangles – the hallmarks of this highly debilitating disease. Yet amazingly, upon examination of their brains, many of the sisters who never showed any signs of Alzheimer's disease throughout their lives also had extensive Beta-amyloid and neurofibrillary damage to their brains.

This proved that people can still function normally even with significant Alzheimer's-related damage to the brain. Analysis of the differences between those who did and did not have symptoms indicated that their writing style earlier in life (taken from essays written when they first joined the order in their twenties) had a large impact on the overall outcome.

Nuns whose sentences were packed with several different concepts were significantly less likely to develop symptoms of Alzheimer's several decades later! Beyond that, another key factor was the mental and physical activity level before and after their daily duties. Those who read avidly and pursued hobbies with great zest in their spare time were still functioning extremely well right into their nineties, whilst those who kept themselves less mentally active in their free time had succumbed to the ravages of dementia.

## Alzheimer's disease versus Alzheimer's dementia

By the time people with Alzheimer's disease-related dementia develop the symptoms of chronic forgetfulness and problems with planning and execution of day-to-day activities, their brains have invariably accumulated a significant degree of damage. The Beta-amyloid plaques and neurofibrillary tangles that appear as dark patches in brain scans were for a long time thought to be what caused the affected brain cells to die off. The latest thinking is that far from being the cause of the trouble, it may in fact herald

an attempt to salvage the dying brain cells. Either way their presence signifies, as far as we currently know, irreversible death of
brain cells.

This irreversible decline usually starts in the Hippocampus
(memory banks – DG and EC stops) and PFC areas vital for the
various mental operations that together fall under the banner of
"Executive Function" – planning, sustained attention, decision
making and the like.

The fantastic news here is that people seem to be able to live a
perfectly good and healthy life even when the unstoppable ravages
of Alzheimer's disease are well under way so long as they have built
up cognitive reserve by regularly challenging their brains.

## What exactly is cognitive reserve?

Two individuals can have exactly the same degree of Beta-amyloid
and neurofibrillary damage, but one experiences a negligible
impact on daily activities whilst the other is dependent on family
or carers to get through the day. How can this be?

Well, it all boils down to what has become known as cognitive
reserve. What sets them apart is, almost always, their level of education and lifestyle choices. More educated people are more likely
to adopt healthy diets, relatively high levels of physical activity and
lifestyles that are consistently mentally stimulating. This in turn is
likely to result from positive childhood associations between physical exercise and mental stimulation that have fostered long-lasting
habits maintained throughout adulthood.

But this doesn't mean that less well-educated people can't also do
this. All they need is the motivation to get stuck into these cogni-
tively demanding activities on a regular basis. Anyone can spend

time reading, dancing, learning a musical instrument or a new language and playing mentally stimulating games.

People whose childhoods were not so idyllic should take strength from the fact that it is never too late to develop new habits. All it takes is daily adherence to sustained, intensive physical and mental exercise over several weeks and eventually new habits will form that perpetuate these behaviours. It just needs to become part of your daily routine.

By consistently challenging your brain it is continuously compelled to create new synapses between brain wires involved in that particular mental activity. This makes subsequent execution of that and any related behaviours more efficient. They can be executed faster, more accurately and with less effort. After many years of tightly interconnected webs of connected brain wire networks you end up with an awful lot of redundancy. That's a good thing. Redundancy in the brain has nothing whatsoever to do with losing your job and everything to do with being able to get by with bits of your brain not pulling their weight properly or even permanently damaged.

## Redundancy in the brain

Now you know that the key to keeping ARCD at bay is building cognitive reserve. The increased connectivity between Prefrontal brain areas involved in orchestrating attention, planning and problem solving enable any mental operation to be achieved in several different ways. These densely interconnected brain areas can then reorganize to take over functions lost when a certain brain area is permanently compromised through dementia, mini-strokes or head injury for that matter. This accounts for why the more physically and mentally active elderly individuals seem to be able to hold onto their marbles for longer.

Every day of your life your Cortex loses 85,000 neurons, that's one a second!

# The Columbo effect

One strange factor that seems to be positively correlated with delayed onset of dementia is dental health. It sounds strange, but the better you look after your teeth the better the chances of your brain serving you for a lifetime free of cognitive decline. It is however more to do with the immune system in general than anything specific to do with your gnashers. Elderly people who undergo major surgery or suffer a succession of minor illnesses are more likely to experience accelerated cognitive decline. It would seem that anything that excessively mobilizes the immune system to aid recovery actually accelerates the process of ARCD.

What put researchers on to this possibility was the fate of the wonderfully talented Peter Falk who famously played the shabby, loveable detective Lieutenant Columbo. At the beginning of 2007 he was suffering from mild cognitive impairment but mentally sharp enough to be at work shooting a film, *American Cowslip*. Sadly, after a series of major dental operations, his cognitive abilities went into a tailspin within just a few weeks.

Another line of scientific investigation also supports the concept that the immune system is very much involved in the development of Alzheimer's. Many middle-aged and elderly people suffer from rheumatoid arthritis. This is where the immune system attacks the joints leading to painful swelling, which in turn eventually causes stiffness of the affected joint. For a long time there was no cure but, in recent times, a new class of drugs called biologics have revolution-ized treatment of this condition by eliminating the compo-nent of the immune response that causes the inflammation in the first place. Curiously, people that have been on this treatment for many years show an extremely low incidence of Alzheimer's dementia compared to others of the same age and from the same socioeconomic background. Could these drugs also be protecting against immune responses that accelerate cognitive decline? No one yet knows for sure, but it is an exciting possibility.

Memories are consolidated in your sleep – the better you sleep the more you will remember – people who nap after memorizing information recall more when tested.

## Chapter takeaways

- You might not be able to stop it but you *can* slow it down. Start applying the brakes now to delay Age-Related Cogni-tive Decline.

- Don't let free radicals become a bull in your china shop. Give healthy nutritious foods packed with antioxidants the upper hand, let them play a regular part in your diet.

- If you want to keep your brain maintained in full working order it's vital you look after the pump that keeps it pumped with everything it needs – your heart.

- Keep your pipes clear. Take the stairs instead of the lift and when possible, leave the car behind, walk or cycle instead. Give smoking a miss and, if you haven't already done so, reduce your intake of red meat.

- Stop the rust by keeping your brain's cogs well-oiled and turning smoothly with mind-stretching, rewarding activities.

- Build cognitive reserve. Whether you're going to learn an instrument, play chess, get up and dance, read, speak other languages, play table tennis or start juggling – do it regularly and keep on doing it for the rest of your life!

# What Next?

You should now be more than aware of just what you have in your possession. You are the owner, guardian and keeper of the most advanced and most sophisticated super bio-computer ever created. How you use it, how you look after it and what you do with it is of course entirely up to you.

## Adrian's reflections

*Three things that have really sunk home for me are the importance of:*

- *continuously keeping my brain challenged*
- *keeping it fuelled with the right fuel*
- *giving serious consideration as to what my gut feelings are telling me*

*(Continued)*

*Feeling mentally sharper and with more consistent energy levels, I am now far better equipped for delivering my motivational talks and workshops. The best news is that when I arrive home from travelling, I'm still up for some serious fun with the kids. And, much to the astonishment of my friends, I'm even getting out on my road bike a lot more. Having already completed my first charity ride I'm now planning to do more adventurous ones later on this year.*

*What has surprised me is just how addicted I had become to checking my emails, text messages and tweets! As a businessman I'd be mad not to still check them regularly, but on an hourly basis rather than every few minutes.*

*In recent years I have become good at boxing off the work from play and have learnt to make time to be with my family and friends. With the knowledge I now have, I am currently focusing on regularly setting aside pure rest time to really wind down and give my brain a chance to catch up with those much needed repair works.*

**– Adrian**

## Jack's reflections

*As a keen sportsman, playing football twice weekly as well as going to the gym, I thought this gave me a licence to eat and drink whatever I wanted. I was forever indulging my impulses by eating fatty and fast-release carbohydrate foods – whenever the fancy took me.*

*Over the course of writing this book I've finally convinced myself to slow down when these impulses take hold and think about my heart, brain and waistline – in that order.*

*A side effect of brain-focused changes to my diet is that I haven't been this lean since my early twenties and I've never had such a capacity to focus without getting distracted for hours at a time.*

*I take a 20-minute nap every single day but not at a set time – I listen to my body. Whenever I've been working flat out for an hour or two and feel myself flagging I get horizontal, set an alarm for 20 minutes later, bring an unresolved problem to mind and go fishing for great ideas. Last month I even finally got around to buying an old school digital wrist watch so I can finally check the time in the middle of the night without looking at my smartphone.*

**– Jack**

# What could possibly hold anyone back?

During childhood your brain cells got wired together according to your own personal experience of the outside world and the people in it. The result is that each and every one of our brains is individually and uniquely moulded by whatever has gone on around us, by what we've seen, heard, touched, tasted, felt and smelt in innumerable different situations.

Young children have a natural, unbridled curiosity which encourages them to set off on a voyage of discovery to help gain a better understanding of the big and exciting world that surrounds them. As soon as a toddler begins to grasp the ability to speak, they'll be off on a relentless quest to find answers to anything and every-thing. With no mental barriers in their way and driven by an unstoppable need to know, their appetite for knowledge, as any parent will tell you, is ferocious. A survey once claimed that a typical four year old asks 437 questions a day – didn't realize it was so few!

The feedback children receive from their constant questioning has a critical impact on their ability to accelerate their understanding of the world around them. Whether it be through information gained from older, more experienced humans, or what they have learnt first-hand through their own trial-and-error exploits. Through both these mechanisms a set of beliefs about how the world works begins to form. Once we have developed a sense of self we start to develop beliefs about ourselves. These self-beliefs start to develop over the course of early childhood with the feedback our brains take on board being key to their formation.

To start with, our beliefs are easily moulded but, as the years roll by, they slowly begin to firm up. Gradually they become less flexible, until in old age they all too often appear to become as solid as reinforced concrete and invariably provide the firm foundations of a deeply entrenched outlook!

*Self-beliefs, whether formed through information obtained from adults or through our own exploratory adventures, are rarely accurate representations of reality.*

## The feedback we received and the impact it had

*"Sticks and stones may break my bones but words will never hurt me"* may have afforded many children over the years some degree of

comfort but, as far as our brains are concerned, this old ditty could not be further from the truth.

It's a sad fact that much of the feedback we received as children was misguided, inaccurate, misinterpreted or just plain wrong. Just a few ill-considered, ill-timed, spur-of-the-moment, negative comments from an influential person can trigger off a nosedive in self-belief. That's because once you start to believe something *confirmation bias* sets in.

Confirmation bias describes a filtering mechanism of your brain that registers and accepts information. It agrees with what you already think and rejects what you don't. So if, for example, based on comments made to you as a child, you have accepted that you are hopeless when it comes to dancing, then you'll have the unfortunate tendency to listen to any future comments that confirm this notion and ignore most of what goes against it.

## Can't hear you!

Just how desperately some people hang on to their existing beliefs was demonstrated when an experiment was carried out with people who held strong political beliefs. They were shown film footage of the leaders of their favourite political party and also their least favoured party. The two leaders were making speeches in which they both totally contradicted themselves. When interviewed afterwards, viewers seemed completely oblivious to the contradictions made by their favoured politician but were highly scathing of the inconsistencies in the opposition party leader's speech. Why let a few simple facts get in the way of a firmly held prejudice!

*(Continued)*

In another experiment where participants were played recordings of people talking about religious beliefs that were in direct contradiction to their own, whenever they were given the opportunity to drown out what they were hearing by turning up an interference noise button – they took it!

# Repetition rules

Should a certain theme crop up again and again, that's when a young person's self-belief really starts getting backed up into a cul-de-sac. Blocked in and dented, they are unlikely to be overly enthusiastic about venturing out of their comfort zone next time round.

When the feedback we received was wrong, we'd either find ourselves believing that we are good at things that in reality we're not (think of some of those poor deluded souls who audition for TV talent shows!) or that we are bad at things that in reality we could become very good at. Thankfully, life usually sets things straight in the former instance but unfortunately not nearly so often in the latter. The vast majority of people think they are useless at certain things that they could possibly be brilliant at. Countless numbers of people go through life blissfully unaware that they have been held back early on in life, never getting the chance to realize their true potential. Simply because certain "grown-ups" had a habit of making repetitive, negative, flippant remarks, they have unwittingly become prisoners of other people's thinking.

More often than not comments are made by adults in an attempt to vent personal frustration rather than to give a fair assessment. Whatever the motive, the long-term effects can be both damaging and profound. Bearing in mind the very people making the comments may well have had their own self-belief knocked and dented somewhere down the line, it's easy to see how a negative cycle can develop. When a person has damaged self-belief, admitting that others are better than them at something can at times be too much to bear and sadly, criticism replaces praise.

The consequence is that, many years later, when any recipient of the inaccurate feedback then comes to perform a certain task, they stumble. That voice in their head says, as it always has said on these occasions – "You've never been any good at this!"

The fact that the personal feedback we received during childhood wasn't always true to reality was, for us as children, completely unavoidable. It obviously varies considerably, being far worse in some childhoods than in others, but whether you realize it or not, we've all had our fair share of inconsistent feedback somewhere along the way.

To be absolutely clear here, the odd comment now and then will not harm a child's long-term prospects – kids are pretty resilient and, as far as adults are concerned, a negative comment here and there might just be the wakeup call needed to motivate them to kick off their slippers and get going. Problems only arise when a certain theme is repeated regularly throughout childhood. Repetition is what causes the cement to set. That's when an adult's harmless comments begin to make an indelible impression on young impressionable minds, flavouring all future experience.

## Photo albums

When we experience anything we never get the full picture all in one go. Instead we take away lots of small, snapshot pictures of different experiences. Think about what you did yesterday, you won't be able to picture the whole day but you will have several pictures of different bits of the day. These snapshots all combine and build up together in your brain to form much larger, jigsaw-style pictures of how the world works and how we fit into it. These big pictures will never, ever be complete because of the scale and complexity of life around us and limitations in our brain's ability to take it all in. There will always be pieces missing. With our brains having nothing else to go on, this is what our beliefs are based on.

What really counts when it comes to forming and re-enforcing self-beliefs is that a certain, constant theme is encountered; in other words, when similar snapshot views are glimpsed over and over again and, the same old familiar pictures keep popping up in our mind.

## Memories are formed of peak experiences and the most recent ones

Would you believe that studies have shown that people actually develop less harrowing memories if an excruciatingly painful medical procedure lasts longer – so long as the most intense peaks of pain and/or the very last experiences of pain are less painful than the shorter procedure? This is because the memory of any experience is formed chiefly from the highlights. However, as the memory of an event is constantly updated, much of the previous information is overwritten by the very last thing. Which is why the ending of a holiday, movie or a painful experience has such a powerful impact.

*Your brain will change according to the consistent messages it receives from the outside world and your self-beliefs will continue to be moulded in a similar fashion.*

Advertising companies are more than aware of this; they know that good branding is all about consistent messages presented in novel, unusual and emotionally-stimulating ways. It has been estimated that, on average, each one of our brains is exposed to up to 5000 ads a day. Most are just a background blur, with only around 1% having any conscious impact. 1% may not sound a lot but that's up to 50 messages a day that your brain is likely to take onboard, at least on a subconscious level. Like it or not, this nonstop bombardment of messages does have an influence on our decision making, that's why companies spend vast sums of money on it. Ad campaigns are most effective when the message a target audience is picking up on is consistent and coming at you from all directions. Yes, re-enforcement is indeed the name of the game.

## Believing is seeing?

Tragically, self-belief for the vast majority of people does tend to stay blocked in.

There are, however, exceptions. They are individuals who have realized that they don't have to watch others achieve great things to realize they can be achieved.

As believers in life before death, they are ordinary people who achieve extraordinary things. They get out there, make the most of whatever talents they have, smash open long-held, restrictive beliefs, push back boundaries, turn current thinking on its head, really make a difference – and then watch others follow.

You have the keys to the world's fastest high-performance engine. If you want to keep on driving up and down the same streets and remain within limits set by others, that's fine. It's your brain and it's your choice. But what a waste!

Hopefully, having read this book you'll be curious to discover more about your own brain, you'll be inspired to want to open up the throttle and find out for yourself just what that precision-tuned, pulsating pink blob between your ears really is capable of!

For information about Adrian – please visit www.adrianwebster.com or tweet @polarbearpirate

For information about Jack – please visit www.drjack.co.uk or tweet @drjacklewis

For references and suggestions for further reading please visit: www.sortyourbrainout.com.

# Acknowledgements

**Jack** would like to thank an awful lot of people. This book is dedicated to his loving parents Phil and Virginia Lewis for providing for an education they never had the opportunity to enjoy themselves, for many fascinating discussions about biology and psychology and for helping me believe deeply both in myself and that anything is possible if you set your mind to it. "So long as you can honestly say to yourself that you have given it your best – then that's good enough for us" was a consistent refrain growing up in the Lewis family household that made high performance easy by taking away all the stress. Points for effort rather than points for achievement was always the spirit of support in our home and it helped me not to panic when the pressure was on in innumerable situations. I am much obliged to my brother and sister, David and Mel, for putting up with their big brother with tolerance and amusement, despite decades of often eccentric behaviour.

I have had many excellent teachers all of whom I wish to thank for imparting their knowledge and a love of acquiring it to me over the years. I would like to thank all the teachers of North Ealing Primary and Middle school for moulding my brain nicely through childhood and those of Latymer Upper School for developing my intellectual, communication and sporting skills throughout adolescence. I have never enjoyed being taught more than during the lessons of my beloved biology teacher Marian Nott – who inspired my passion for the brain by strategically placing a magazine about neuroscience at the front of the class for me to find.

All the tutors that lectured me during my Neuroscience BSc at the University of Nottingham deserve high praise, in particular Dr Kevin Fone and Prof. Charles Marsden (whom we undergraduates often referred to as God based on his seeming omniscience). Of my many undergraduate neuroscience colleagues who never tired of putting up with my hare-brained ideas, nor begrudged me their lecture notes when it came to revision time – I must give special mention to Dr Kate Atherton, Dr John Messenger and Oxford University's Dr Andy Sharott. My dear friends of the OTM Collective deserve special mention for endlessly distracting my attention with flyering and flyposting duties. To build a successful club night from scratch whilst simultaneously earning a 1st class degree in neuroscience was no mean feat. Thanks for the challenge.

I owe a great debt of gratitude to Prof. Semir Zeki for taking me on to do my PhD at University College London and teaching me what being a scholar is all about; without doubt the most intellectually stimulating (and intimidating) experience of my life so far. To Dr Matt Self and Prof. Andreas Bartels who taught me everything I know about Matlab coding and fMRI studies. To "opt out" Ollie Hulme (PhD), Barrold Roulston (PhD), Ulrich Kirk (PhD) and all the other "baboons" of the Zeki lab for their help, moral support and for accompanying me to SOAS on so many great Friday nights for endlessly inspirational discussions about brains et al.

I would like to thank all the principal investigators at the FIL in Queen Square for opening my mind to the broad reaches of neuroscience research and for invaluable feedback on my studies. A special mention must go to my second supervisor Prof. Karl Friston (quite the loveliest and most intellectually stupefying person I have ever met), but also Prof. Ray Dolan, Prof. Cathy Price, Prof. Jon Driver (RIP), Profs. Chris and Uta Frith who continue to inspire me to this day. And on to my post-doctoral studies at the Max Planck

Institute for Biological Cybernetics in Germany. Thank you so much Prof. Uta Noppeney (now at University of Birmingham) for forgiving me my three-year excursion into television and having enough faith in my potential to bring me into your lab and provide the foundation upon which our Journal of Neuroscience paper could be realized.

I owe a great deal of debt to my colleagues Dr Sebastian Werner and Dr Hwee-Ling Lee for their tireless counsel, as well as Prof. Heinrick Buelthoff, Prof. Mark Ernst, Dr Chandrasakar Pammi and many others for their help and support. Thanks also to Andi Lampi and the Roigel for keeping me sane in those months of isolation on the edge of the Black Forest. In order for these acknowledgements to not exceed the length of the whole book I will conclude by thanking everyone who has ever encouraged me to pursue a career in neuroscience outside of mainstream academia.

My good friends Scott, George, Xan, Ben, Dan, Festa, Leo, Eco, Laz, Vladimir, Chris, Chris, Henry, Angie, Tammy, Emmy, Amy, Sophie, Jess, Emma, Geordie Chris, Def and many others – your encouragement and support kept me going through times of despair and re-kindled my determination to pull off yet another unlikely achievement. And finally to three very special women, who each nourished my soul for a period of several years along the way, to whom I owe a special debt of gratitude: Kate, Bex and Lucie.

**Adrian** would like to thank James Poole for being such a great friend.

Adrian and Jack would both like to thank Charlie for his wonderful illustrations, Neil Crespin for bringing them together on that fateful day in Tenerife and Holly Bennion whose idea it was that they write this book in the first place.

# About the Authors

## Dr Jack Lewis

Dr Jack Lewis has a first class neuroscience BSc from the University of Nottingham and a PhD in neurobiology from UCL. His postdoctoral research, conducted at Germany's Max Planck Institute for Biological Cybernetics, explored multisensory perception using fMRI. Regular speaking engagements enable Jack to motivate and inspire school, business, festival and conference audiences with pearls of wisdom fished from the neuroscience literature. Jack has made neuroscience accessible, compelling and (hopefully) useful for global audiences of millions through television programmes for BBC, ITV, Channel 4, Channel 5, Sky, Discovery, National Geographic Channel, CBC, TLC and even MTV. Jack is MD of the consultancy firm Neuroformed Ltd (www.neuroformed.com).

For information about Jack – please visit www.drjack.co.uk or tweet @drjacklewis

# Adrian Webster

Milkman, policeman and salesman were just a few of the entries on Adrian's CV before he moved into the IT industry and discovered an ability to inspire everyday people to achieve extraordinary results.

The son of a Yorkshire coalminer, he is now an international bestselling business author and one of the most popular motivational speakers in Europe today – delivering keynote presentations around the world.

For more information regarding Adrian's speaking or his workshops, please visit www.adrianwebster.com or tweet @polarbearpirate

# Index

*Index compiled by Annette Musker*